MW00682511

ON LEAVING BAI DI CHENG

Let us unite as one,
against the great gusts of wind!

For the preservation of the
Culture of the Yangzi Gorges

If you want to help . . . For more information on how you can help save Chinese cultural relics that will be flooded, write to The Ontario Archaeological Society, 126 Willowdale Avenue, North York, Ontario, Canada M2N 4Y2. Letters will be forwarded to the authors.

CAROLINE WALKER RUTH LOR MALLOY
ROBERT SHIPLEY FU KAILIN

ON LEAVING BAI DI CHENG
THE CULTURE OF CHINA'S
YANGZI GORGES

NC Press Limited

Toronto 1993

Photo Credits
Front Cover:
    Badong will be flooded by the three Gorges Dam reservoir. (Shipley)
Back Cover:
    Top Left: Tujia tracker. (Lor Malloy)
    Top Right: Big Buddha at Chongqing. (Walker)
    Bottom Left: Hauling boats against the current. (Walker)
    Bottom Right: This Han pillar will be moved. (Walker)

©Caroline Walker, Robert Shipley, Ruth Lor Malloy, Fu Kailin, 1993

No part of this publication may be reproduced, stored in a retrieval system, or transmitted, in any form or by any means, electronic, mechanical, photocopying, recording or otherwise, without the prior written permission of NC Press Limited.

Canadian Cataloguing in Publication Data

Main Entry under title:
    On leaving Baidicheng: the culture of China's Yangzi Gorges

Includes bibliographical references and index.
ISBN 1-055021-083-1

1. Yangtze River Gorges (China) – Antiquities –
Collection and preservation. 2. Dams – China –
Yangtze River Gorges. I. Walker, Caroline (Caroline Marie), 1942-

DS793.Y305   1993   915.1'2   C93-095244-8

We would like to thank the Ontario Arts Council, the Ontario Publishing Centre, the Ontario Ministry of Culture, Tourism and Recreation, the Canada Council and the Government of Canada, Department of Communications, for their assistance in the production of this book.

New Canada Publications, a division of NC Press Limited,
Box 452, Station A, Toronto, Ontario, Canada, M5W 1H8.

Printed and bound in Canada

Contents

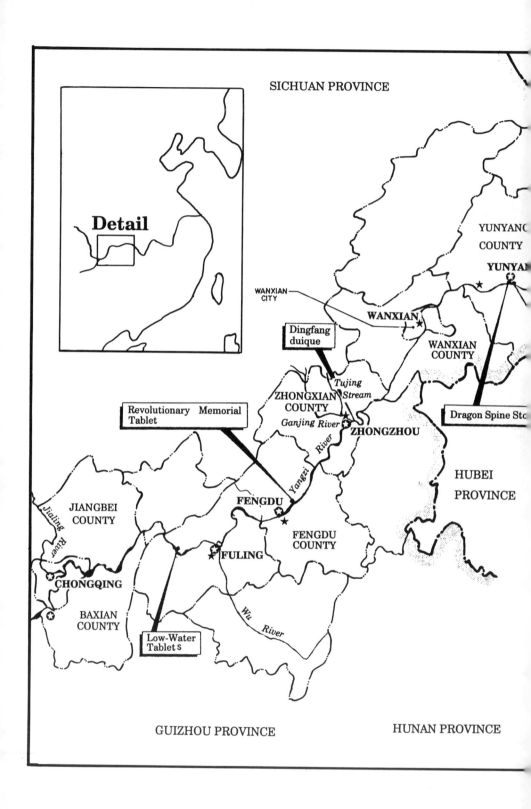

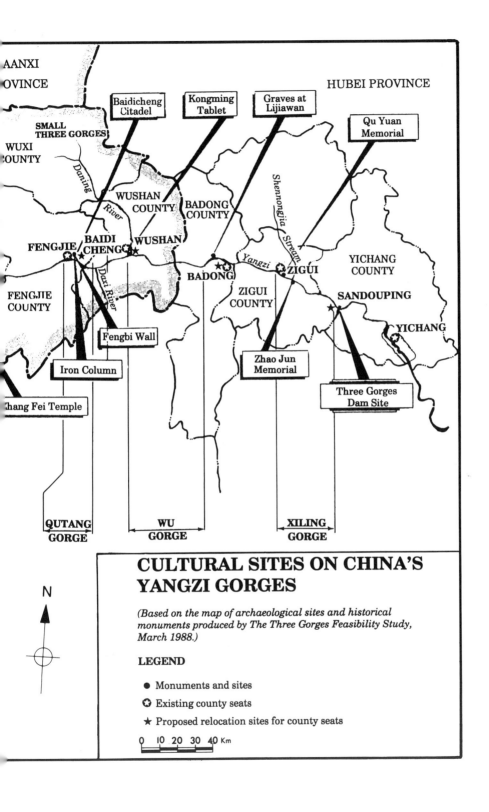

SHAANXI
PROVINCE

HUBEI PROVINCE

SMALL
THREE GORGES

WUXI
COUNTY

Baidicheng
Citadel

Kongming
Tablet

Graves at
Lijiawan

Qu Yuan
Memorial

Daning River

WUSHAN
COUNTY

BADONG
COUNTY

Shennongjia Stream

FENGJIE

BAIDI
CHENG

WUSHAN

BADONG

Yangzi

ZIGUI

YICHANG
COUNTY

FENGJIE
COUNTY

Daxi River

ZIGUI
COUNTY

SANDOUPING

YICHANG

Fengbi Wall

Iron Column

Zhao Jun
Memorial

Three Gorges
Dam Site

Zhang Fei Temple

QUTANG
GORGE

WU
GORGE

XILING
GORGE

N

CULTURAL SITES ON CHINA'S YANGZI GORGES

(Based on the map of archaeological sites and historical monuments produced by The Three Gorges Feasibility Study, March 1988.)

LEGEND

● Monuments and sites

✪ Existing county seats

★ Proposed relocation sites for county seats

0   10 20 30 40 Km

Acknowledgements

Our very great thanks to Wang Bin, Nie Shijia, Zhen Xianlong, Wang Changchun and Tian Yuan who took time out of their busy days to show us the wonders of the Gorges.

I would like to thank my husband, Norman Endicott, for bringing me to China, and my sister, Janet Walker, and my mother, Jean Walker, for reading the manuscript and giving their good advice. For all of their help, translating, explaining and teaching me about China, as we went in pursuit of the Neolithic, Chen Shen and Bing Wang, thanks. Thanks also to Geng Jie, Geng Li, Lan Renzhe, Wang Yu, the Huangs and Francoise Pinot. *Caroline Walker*

In the past and present, my mother has, and continues, to support me in ways that make all things possible. Norman Endicott and Caroline Walker gave me the opportunity to go to China. Huang Zhongling and Fu Kailin shared their love of China with me and helped me to learn many things. Among the gifts my wife Pam gives, is help in allowing me to appear literate. Thanks to all. *Robert Shipley*

Thanks to David Barrett, Mike Conrad, Yue Chi, John Ma Feng Jian of China Shandong Tourism Corporation, Jinan, China. Also the East Asian Library, Royal Ontario Museum, Holiday Inn Crowne Plaza and China World Hotel. *Ruth Lor Malloy*

Thanks to Prof. Zhou Guangya, Prof. Lin Xiang, Prof. Prof. Ma Jixian, Prof Chang Qunli, Prof. Xu Xiafeng, and Prof. Xia Wenrong. Also thanks to my family that helps me to do interesting things. *Fu Kailin*

1

CONVERGING STREAMS

On Leaving Baidicheng

(Robert, Conestogo, Ontario, October 25, 1992) The phone rings one Sunday afternoon about a month after we return from China. It is Caroline Walker. "We need a title for this book because next year's publishing catalogue is being prepared," she says. "Phone me back in about ten minutes and give me a suggestion." I go upstairs to the study and open the notebook I kept on the trip down the Yangzi and get out a couple of fans I had been given by one of our Chinese friends. These are carefully inscribed with poems in classical calligraphy. "Ten minutes, eh."

But it does not take that long. Poetry, it seems to me, is how people of most cultures and languages have done their best at capturing broad sweeps of emotion and substance with verbal economy. When you are inclined to carve the essence of your thought into solid rock in complicated written characters, as the Chinese have been wont to do all along the walls of the Three Gorges, economy of form is necessary. On page 118 of my China notebook I had glued a scrap of flimsy newsprint torn from a tourist brochure. On it is printed, in an artistic form of Chinese calligraphy, a couple of dozen characters. On the opposite page is a translation that had been rendered by one of our Chinese travelling companions, Huang Zhongling.

> *It is like coloured clothes in the sky*
> *In the morning when I leave Baidicheng;*
> *We need only one day to go to Janglin . . .*
> *One thousand li.*
> *There are so many monkeys crying on both sides of the river;*
> *My fast boat has already gone through many mountains.*

"On Leaving Baidicheng," by the eighth century Tang Dynasty poet Li Bai, or Li Po as his name was represented years ago when I studied his work in university, is one of the most famous pieces in all Chinese literature. The cultural significance of this fragment of verse can be gauged by the fact that a man like Zhongling, who has no pretence of being a scholar, would not only be able to translate it for me but would take considerable pleasure in doing so. My old university text, I might add, features a far more prosaic rendition.

But to appreciate fully the significance of Baidicheng (White King City) a more complete picture needs to be drawn. The place is roughly half-way along the six hundred kilometre stretch of the Yangzi River that Caroline and I travelled with three Chinese friends. It is about half-way between the cities of Chongqing and Yichang. This is the area that will be flooded when the largest power project in the world is finished.

We arrived at the site on the seventh day of our 12-day trip on the river. After a tortuous half-hour ride over a typically unpaved mountain road in a bus with poor suspension and without padded seats we had come to a very inauspicious looking coal-mining village. It was hot. Far below us were the muddy waters of the Yangzi. Stretching up from the village were the proverbial "ten thousand steps." There were sedan chairs for hire, each carried between poles on the shoulders of two men. We refused the luxury. We climbed and rested, climbed and sweated and climbed again until we reached what seemed like temple gates. What we found was a kind of combination of museum, historic site and interpretive centre. The collection of buildings, mostly built in the 1920s, had in fact been the villa of the Guomindang poet-general Wu Peifu. The warlord gratefully donated the mountaintop retreat to the state after the defeat of the Nationalists by the Communist Party in 1949. There was a collection of artifacts including calligraphy carved in stone tablets, and dioramas with life-sized statues of historical and mythological figures.

I was not prepared, however, for what I beheld when I emerged from the far side of the complex. Baidi is a solid little mountain that must rise about five hundred metres above the river. It stands opposite Kuimen, the gate or entrance to the first of the Three Gorges. There the mountains tower almost vertically another thousand metres straight up. Far below, the swirling brown waters of the Yangzi are swallowed in a cleft between these magnificent rocks. The gap cannot be a hundred metres across. At the moment I saw this scene, through the salty sting of the sweat in my eyes, the sun was so bright that one side of the gorge seemed to be gold and the other side black as coal.

In my journal I noted that I had "seen a lot of wonderful and beautiful sights in my life — Mont St. Michel, Lake Louise, Massada at sunrise, Santorini, et cetera. But when I saw this place, I must record, tears came to my eyes."

We spent that night in the beautiful guesthouse whose upturned eaves stand high above the orange groves on the shoulders of Baidicheng. From the courtyard one could count the stars, crisp and brilliant in the absence of light pollution. This mountaintop and the nearby slopes have been garrisoned at least since the Han Dynasty. From here all the traffic on the river could be controlled and taxed. In the afternoon we had sifted through the remnants of a two-thousand-year-old city wall, carved into tillable terraces by generations of patient peasants. And here, too, the poets had been drawn by the beauty and mystical quality of the place. Du Fu, Li Bai's Tang contemporary, had lived in a villa just up the valley behind Baidicheng. When Zhongling returned to our room from his own solitary nocturne, he said he had sat wondering what it would have been like to have been here in ancient times.

I noticed much later in one of the photographs I had taken that the spectacular view of the Gorge was obscured by some electrical wires. It reminded me that in the name of modernization and power production this scene is about to be changed forever. One of the most famous places in the history and literature of China will no longer be as the ancient poets found it and left it.

If the Power and the Fax are On

(Caroline, Chongqing, September 16, 1992) Six a.m. The alarm awakens me, for the tenth time: pitch dark. Our papers are the last thing to be sorted. What do we really need? The report on the archaeological sites in the Gorges, maps, tourist brochures. Nothing heavy, nothing controversial. I will carry my gray knapsack and the camera case. The video camera and film are heavy, and it will only be good for a couple of hours if I am unable to recharge the batteries.

Robert has most of the still film and the tripod. The tripod is light, but its case is already stuffed with Zhongling's jacket. His camera bag, with two camera bodies and several lenses, is heavy. With one bag each for our Chinese friends, not counting purses, there will be ten or eleven pieces to start. Li will never offer to be a porter for these crazy foreigners. Her knapsack does not seem to weigh her down: she is the smallest of us (and the youngest) and she always strides on ahead.

Already I have a plastic shopping bag holding a bottle of water, Norman's jacket for me, and a carton of Marlboro cigarettes Robert brought to give away. Repeatedly Robert and I will ask, "Shall we give this guy cigarettes?" They are not heavy but they take room. Zhongling, our expert in such things, keeps saying, "Money is enough." Cigarettes in modern China have a complex etiquette and imported cigarettes have the most prestige but, perhaps today, cigarettes are not enough. The bag also contains several English issues of *China Daily*, the international version of articles from Chinese newspapers. I have not read them but mostly I am bringing them because I am an old China hand: they will be needed to sit on.

My husband Norman gets up to have breakfast with us. The hotel coffee shop is full of German tourists, leaving on a luxury cruiser. Afraid that the staff won't have time to serve us, I leave Norman to deal with them. The tall blond foreigner, who was born in Chongqing and speaks Sichuanese, has become quite a beloved character around the hotel.

In 1988 Norman and I, his son Eric and his grandson Paul, took the Three Gorges tourist cruise from Chongqing to Yichang. The ship was nice enough, although the chef was not very good. We were on our own. The guidebooks and on-board guides did not have much to say and what they did say did not interest me much. We entered the first gorge in early morning darkness and after that, well, I had seen a lot of mountains and these were not high mountains. The trip in the small boat up the Daning River was wonderful and so, all and all, the trip was satisfactory. But I had not wanted to do it again. Until now.

Chongqing became Norman's and my home for a year, in 1990. Its frontier energy and can-do spirit appealed to us in 1988, and, besides, Norman was born there in 1926, the eldest son of the Rev. James and Mary Endicott, missionaries for the new United Church of Canada. James Endicott had been born in Leshan (Jiading), a small town in central Sichuan, in 1898, and so Norman is, as he proudly says, the third generation of his family to have lived in Sichuan.

James Endicott eventually became a teacher at the Gini *(Jinling)* Middle School, and he wrote a series of books for teaching English under his Chinese name, Wen Yuzhang, that were widely used in China. In 1938 he went to work for Chiang Kaishek, helping in the relief of civilians during the Japanese bombing. In 1943 he became a teacher at the West China University in Sichuan's capital, Chengdu, where he was active in the struggle for democracy and an end to the civil war that broke out when the Japanese were defeated. In 1947 he moved to Shanghai, where he published an underground newspaper. He returned to Canada and became a tireless advocate of peace during the Cold War. Although he has not always agreed with them, he has been honoured many times by the Chinese government.

The Canadian economy was sliding into recession and riding out the bad times of 1982 at the helm of a small publishing company had been no fun for me. Parkinson's Disease had put an end to Norman's law practice in 1989. The Chinese cultural attache in Toronto had said I should forget about getting a job in China: no PhD. But we knew that the secret to success in China is "connections," *guanxi,* and the Endicotts have that. Tiananmen was in our favour: who wanted to go to China in the winter of 1989? Teachers were needed, even small business-owners with enthusiasm but no teacher training.

Norman's Chinese name, Wen Xiaozhang, is so similar to his father's that, as a dutiful daughter-in-law supporting his family in the Chinese tradition, I found many doors open to me. On Christmas Day, I got a phone call at 7:00 a.m. from the new president of Sichuan International Studies University (SISU), Lan Renzhe. Can you come to teach English by February 10? Why not! On December 29, 1989 I had an emergency hernia operation. We went. It was the best year of our lives.

I sought out the Chongqing Municipal Museum, and then introduced some of my colleagues and students to it. The curators and officials at the museum were glad to be "discovered" as museums in China want to reach out to the outside world too. They helped us, officially and unofficially, to see some of the local historic sites that are not on the regular tourist agenda in Chongqing, and my interest in local history was piqued. Everything was Ba: a tribal people who had made Chongqing their capital 3,000 years ago. Young artists were drawing on what they took to be the Ba style for wooden masks, batik, fabric design for fashion, painting, and sculpture. I also met local historians who had paid dearly for going against the grain in the various campaigns and purges. They showed me their photographs and lifetime of hand-written manuscripts. They believed that their area of the world had its own story, and that now they would have their chance to tell it.

When the Chinese Peoples' Congress passed the enabling legislation for the Three Gorges Dam to go ahead, everything came together: the archaeology courses I had been taking at the University of Toronto since my return in 1991; our continuing ties with our friends in Chongqing; and the opportunity to do something interesting and, I hoped, important. The archaeology books on China available in English had almost nothing on southwestern China. The materials I had been able to gather on the Three Gorges were frustratingly vague. And yet the area was one of the cradles of China's magnificent civilization.

And so the idea for our Three Gorges Project was born. Robert Shipley was the author of a book NC Press had published on Canada's war monuments, and he was back at Waterloo University finishing his M.A. in heritage planning and conservation. He agreed to join Norman and me on our trip to Chongqing in September 1992, to go through the Gorges — but not the tourist way. Ruth Lor Malloy was a China guidebook writer with a passion for handwork, "women's art," who had taught me much about China. She could add the Gorges to her 1992 itinerary in October, to go to the places we would miss.

The "China" connection was more difficult. Communication with Toronto is getting better, but it is still slow and haphazard. Rather, the mail is slow (ten days each way) and the phone and fax are haphazard, dependent on getting through at workplaces and the power and the fax machine being on. And so while plans to have some Chinese friends accompany us had been in the works for months, the composition of our group was in doubt almost to the last moment.

I met Fu Kailin through Norman. He had been treated by a Dr. Luo, and Kailin was Luo's friend and translator. When I first met her I thought her intensity unusual in a Chinese intellectual, and a bit overwhelming. I came to love her for it. There can be no truer friend, and her enthusiasm for new interests and projects (so they say) matches my own. A teacher of English at the Teacher's College in the Chongqing district of Shapingba and a director of a Fine Arts Institute, she has a great and wide network of friends in Chongqing. It was Kailin who introduced me to the Ba and to those who studied their culture. One of our friends was Li, the only one of us trained in Chinese archaeology. Another was Francoise Pinot, a French teacher at SISU, who could not leave her work to come with us on this trip. Zhongling, our "Chinese nephew," works for China's coal monopoly. We shared a passion for Chongqing history.

After a year studying Chinese archaeology, I had my own reasons for my interest in the Gorges. Our project was received with scepticism by some of my Chinese friends: A scenic area, but nothing important happened there. And one of my students had bitterly told me of being called a "Sichuan rat" when she travelled to Shanghai. Sichuan people are short and dark and move quickly and are said to be too clever by half. This prejudice, north against south, was borne out in the archaeological literature until recently.

A popular theory says cultural change occurs in "Nuclear Areas" and then radiates out to the hinterland. A Hong Kong archaeologist, William Meacham, had written (and I had remembered) that South China was considered a hinterland, a "rather simple culture" with an indolent, primitive and backward population under the constant impact of advances occurring in the North. In fact, until recently, "Chinese civilization was traced to an origin in the West, rice agriculture and painted pottery in Asia to North China, bronze in Yunnan and Thailand to Zhou [Dynasty], agriculture in all of Asia to China . . ." (Meacham 1985:90). Meacham argues passionately for local evolution, the potential of the relationship between people and their environment, their previous cultural history and the creativity of each human being. That, I thought, was a good starting point for our investigation of the Gorges area.

For better or worse, our little group had decided not to make a formal application to the authorities to be shown the cultural relics of the Three Gorges. We did not merit any official generosity, and we certainly did not want to face the problems of being "journalists" in a China nervous about reporters on the controversial Three Gorges project. We were afraid that we might run into restrictions, obligations, or special costs. Our time was short and our financial resources were limited to our own. We just wanted to go and see what we could see for ourselves.

Guanxi: Good If You Have It

(Ruth, Beijing, October 24, 1992) I was born in Canada into one of two Chinese families in small, conservative Brockville, Ontario. We were near the bottom of the social ladder, the object of racial slurs. I hated it. When I went to college in Toronto, I suddenly discovered that China was actually much admired: major exhibits in the Royal Ontario Museum and important university departments were devoted to its study. Curious, I decided to visit that country. I finally succeeded in 1965; I saw my father's village birth-place, met relatives and their friends, and got my first look at the Forbidden City and Great Wall. With my different perspective on the then-regimented "Red China" I was able to publish two articles in the national *Weekend Magazine.*

I went to China again in 1972, the year President Nixon made his historic visit. By that time, there were other North Americans in the hotels. I found myself telling fellow travellers how to get around the cities. And I realised there was a need for a guidebook, for a bridge between the East and the West. So I wrote and published one.

There was a need also for improving Chinese hotels and travel agen-cies and transportation. There still is a need, even though in the last twenty-five years magnificent palace hotels rivalling anything in Toronto or New York City have been built. Foreign managers have trained local staff to international standards. But China is a big country. Some of the travel agencies still need training in western-style efficiency. Then there are the trains, and the airlines, and the airports, and the ferries . . . though there has been much improvement here too.

I now keep visiting China and writing my guidebooks and using my position as a writer to drop hints for change on old, atrophied male officials. I like to support the growing number of young managers who have come up through the ranks and know first-hand the wants of foreign tourists. I try to teach better organization, western-style efficiency and service. Some-times it helps just by writing about the problems. And I must admit, as a woman, a gender looked down on by China's male-dominated society, I take much pleasure in asserting myself to authorities in China.

Of course I also keep returning to China because I like the country, its challenges, and its beauty. I like learning about its culture and history; there is no better way to learn than to write (or to teach). I like seeing the changes. When I first went there, foreigners had to travel with a guide. Now one can get around on one's own quite easily, especially in and between the big cities.

In 1992 I visited China to find out if there is anything to be done about the ancient buildings and relics doomed to be drowned by the new Three Gorges Dam on the Yangzi River.

In China, one is advised to approach important people with an introduction. To get to talk to the government person in charge of relics in the area threatened by the dam, I went through Tom Haney, an American working for the Shanghai Centre in Shanghai. The Centre was co-sponsoring a sale of antiques with the State Bureau of Cultural Relics. I happened to be covering it for my publications.

To tweak the Bureau's interest, I first talked with Miss Shi Yazhu of the Bureau's Foreign Affairs Department about a North American tour of an exhibition of imperial Chinese costumes. Miss Shi happened to be in Shanghai for the antique sale. The tour was a genuine, though remote possibility for Toronto's Museum for Textiles, for which I was a volunteer. Miss Shi arranged for an interview with a Mr. Wang Jun during my stay in Beijing.

This networking is important because many Chinese people are afraid of making mistakes, of taking responsibility. Miss Shi would get in a great deal of difficulty if later I was caught stealing relics or pilfering funds. They could now trace me back to Tom at the Shanghai Centre to whom I was also responsible. *Guanxi*, or connections, is the way things work. There is also an element of you-help-me I-help-you, all wordlessly understood. *Guanxi* is a good system — unless you don't have any. The Chinese do not have a monopoly on *guanxi*, but they have developed it into an art.

Mr. Wang seemed pleased with our suggestion to help raise money to save the relics. Money is power in China these days. Still, I cautiously volunteered the name of my contact at the National Tourism Administration in Beijing and watched Mr. Wang scribble down the telephone number while denying that it was necessary. I had been writing travel guides on China for over fifteen years then and was considered in some circles a *lao pengyu*, an old friend.

Mr. Wang looked younger than I had expected; I guessed he was in his thirties. He was dressed in a fancy silk bomber jacket, last year's fashion. He also wore a polo shirt and blue jeans. I had never seen a government official wearing jeans before. I wondered about his rank but he had no business card. What did that mean? Everybody important in China has a business card. Maybe Mr. Wang had genuinely just run out. Still, he did not write his title into my notebook even when asked. Of course I was sure Mr. Wang worked for the bureau. After all, he had been introduced through Miss Shi, who had been introduced through Mr. Haney, who had been introduced through Carole Goldsmith of Shanghai's Portman Shangri-La Hotel, whom I had known for years. Some things you have to take on trust.

Mr. Wang was not the usual government official. He said he was an archaeologist who had been doing field work in the Yangzi area since 1984, sleeping in farm houses. Officials do not usually sleep in farm houses; most have a per diem so they sleep in hotels. I liked Mr. Wang because he was a genuine archaeologist, not a bureaucrat. Mr. Wang explained through an interpreter that the Cultural Relics Bureaus of the two affected provinces,

Hubei and Sichuan, would be making recommendations to his office about the relics. They would be studying what should be saved and how much it would cost. There were eight hundred sites threatened with inundation. The provincial reports would be ready five months later, in March, 1993, he said. He would have appraisals ready by April or May.

I asked if the Bureau wanted help from abroad to save the soon-to-be flooded relics. Yes, it needed assistance to get worldwide publicity. "It is the responsibility of the whole world to care about this project," Mr. Wang said. China needed more advanced technology than it had, he continued. It needed archaeologists and technical experts on the protection of cultural sites. Mr. Wang knew about the many foreigners who had helped at the Aswan Dam project in Egypt. The form of assistance might be different, he said cautiously, but assistance was needed. The government could not pay for everything. He guessed as a total figure: one or two billion Chinese yuan, or about US$182 million to $364 million to save all the sites. (Chinese yuan was calculated at US$1 = ¥5.2. It is now 5.7 on the official exchange market.)

But would the raising of money abroad only be a way for officials like Wang to line their own pockets? The office of the Cultural Relics Bureau was near the northeast corner of the Forbidden City, the famous 15th century Imperial Palace. Outside the wide three-storey red brick building was a row of book and magazine stalls. Around the corner was a block of fruit stalls, lush and colourful with peaches, oranges, apples, and persimmons. To the south, along the east Imperial Palace wall, was a lively morning vegetable and meat market. Almost everywhere one looked, were lines and lines of clothing stalls. Some of the jeans and jackets had American and Italian designer labels. Many of the stalls were privately owned, run by people who could not get such a lucrative income elsewhere, people trying to supplement meager government salaries, people eager to give capitalism a try. The peddlers made considerably more money than government officials like Mr. Wang. Still, I felt I could trust him.

In China, one learns quickly to double-check and triple-check answers to questions. If you question ten people, no matter how knowledgeable, you get ten different answers. People want to be helpful, to answer questions. But they might not have the information, so they guess. Beijing is usually the final word, but one needs to continue asking. Later, a museum director explained an example of the help needed. He talked of the moving of a three-hundred-metre-long stone marker in Sichuan Province. China did not have the technology to do this now. Officials said that all levels of government would contribute to saving the relics: county, provincial and national. One official said that citizens would be taxed the same way as the citizens of Beijing were for the Asian Games: directly through their work units. Considering that the average monthly income of a university professor is ¥350, or roughly US$64, I shuddered to think what this would mean to individual Chinese when they had so many other obligations.

Beijing in late October was dry. I woke up every morning with a sore throat. There is a perennial water shortage here and the government has announced a plan to build or use a thousand-kilometre canal to take water from the Yellow and Yangzi Rivers to service Beijing, Tianjin, and areas like Jiangsu Province along the way. Part of the old Grand Canal, completed in the 7th century AD, would be used. But Beijing also has dozens of fancy hotels, some offering fresh Canadian lobster, California sushi rolls, and shrimp flown in three times a week from the Gulf of Thailand. The hotels have simple-to-use USA-Direct telephones, and eager taxis at the door. There is never any shortage of water in hotels (which wastes much of it because of the flush toilets).

Caroline and Robert had visited all the affected sites along the Yangzi River except for Badong, Zigui and the Yichang area. These were my assignment and so I took a train to Yichang. The Beijing train station at midnight was carpetted wall-to-wall with human bodies, poor people clinging to meager belongings. Beijing has a "floating" population of two million people, who have left their roots in the countryside to try for a better life in the big, wealthy capital. For these poor, a visit to the fancy hotels, to any hotel, was inconceivable. Maria Yang and I dragged my bags through the bodies, rolling over legs, and hitting shoulders; there was no other way. The path between people was narrower than my suitcase and most of them were immobile, asleep. No one objected.

Surprisingly, we could not find a porter until we stood waiting at the gate above my train. The man was dirty, in some sort of uniform, eager to grab my bags before the gate opened, before we could follow him to the train. I did not trust the gleam in his eyes nor the grin on his unshaven face; I imagined my bags disappearing forever. In spite of his persistence, we continued to haul the bags ourselves, even though Maria said, "His heart is good but he does not know how to present himself." Maybe she was right, but I did not want the risk.

In contrast to the filthy station, "soft class" was clean, four people to a sleeping compartment that could seat six comfortably. The bodies in the station would end up in "hard class" seats, in what look like cattle cars, on three layers of berths, or in the aisles (unless they were just treating the station as a hotel). There had been no flights from Beijing to Yichang. That Yangzi River port is 38km from the new dam site, and is the headquarters for its construction. To fly would have meant going first to Guangzhou, Changsha or Wuhan and then taking a twice- or thrice-a-week flight. The promised "overnight" train trip had seemed more appealing.

My trip to Yichang from Beijing had been on-off-on. Maria represented a travel service named BYTS, but BYTS had not bothered to find out until 12 hours before that there was only one express train every two days and we had to go tonight, not tomorrow. It was this kind of sloppiness that kept bringing me back to China (though I am not really a masochist—I really want to help). The Yangzi Valley near the new dam has no big modern cities. I felt I was going back in time to my early days in China, though there would not have been dirt in the train then. Some things have gotten worse. Maybe it is the lack of Communist discipline.

The express train was special. Service had started in 1984 because of the earlier Gezhouba Dam built in the Yichang area and the subsequent push for economic development in the Yangzi Valley. At that time, Yichang was touted as the capital of the new province of the Three Gorges. The new province never developed and no one spoke of it now. Recently an economic development zone has been formed in Sichuan to help with relocating people and development. It will hasten the construction of the infrastructure (railways, highways), improve communications, water conservation, and public utilities. The Yangzi Valley still has a long way to go to catch up economically with the more prosperous coastal areas.

Maria had said Yichang was an "overnight trip." I left at midnight and it turned out to be one-and-a-half overnights. Another surprise. We arrived in Yichang at 3:15 a.m. the second morning. At least it was better than the 48-hour train trip I once took after being promised only 24 hours.

The toilet room was not only filthy, the window was broken and glass slivers lined the floor beside the toilet hole. Maybe someone threw in a stone from outside as they do occasionally at trains and cars in America. I wondered if the neglect was because nobody important travelled to the underdeveloped Yangzi. Staff was there to clean it up, but the glass was not removed during the whole trip. Back in my berth, I secured my purse and camera between my legs under the blanket. Sleep came easily in spite of the strangers in the compartment. I trusted them: a factory official and a kind older couple who had given me an apple. They did not look like petty thieves, and I was right.

Chongqing is Boom City

(Caroline, Chongqing, September 14, 1992) Prof. Dong says Chongqing is the tail of the dragon. (Its head is in Shanghai.) Tax for a joint venture is only 15% (the magic number from Hong Kong): the first two years are tax free. The numbers are danced before our eyes. They vary from place to place, deal to deal. Things are changing very quickly indeed. Dong is launching a magazine to publicize the Three Gorges. The support of the municipal government is already assured, and four hundred copies of a document with the government's red seal have gone out to all areas of Chongqing. My name is on it. There was no time to get my approval of the plan. A foreigner's name gives it credibility.

Later, Norman and I are sitting at a table in the bar at the Chongqing Guesthouse with a friend and a Mr. Xu from the Office of Economic Cooperation of the Chongqing People's Government and chief manager of the Chongqing Economic Cooperation Corporation (CECC). If Chongqing had a Board of Trade, Mr. Xu might well be head of it. The CECC is one of a bewildering number of overlapping companies and government departments, organized almost daily it seems, to facilitate the privatisation of practically everything, in this, the fall of 1992.

CECC was set up in 1988 in a 28-floor office tower. Annual sales are ¥100 million and it has 19 subsidiaries in everything from real estate to training and technology. It is primarily a broker of surplus raw materials (coal, oil, wood) and Mr. Xu has a network of friends all over China. Recently, he lets slip, he shipped some iron ore to Beijing for processing, because the cost is lower there, and then resold it in Chongqing.

Since Deng Xiaoping's trip to Shenzhen last spring, Chongqing's leaders have been changing their minds fast. In 1991 there were 18 joint ventures: as we speak, 92. By 1995 there will be 2000. No matter how little, we are welcome to invest. Two areas have been set aside for a duty-free area that will challenge the miracle of Pudong (in Shanghai). Chongqing's drawback has always been location. Labour and materials are cheap but transportation is the killer. A ¥9 million container port has been in the works for years. Xu assures us that the government will allow foreign investment in transportation: air, rail, shipping, even subways.

Xu has no doubt that Chongqing will be the planning, organizational and financial centre for the Three Gorges Dam. Li Peng heads the project so that means it will go ahead. Construction will bring years of increasing opportunity for Chongqing. Hundreds of thousands of people will be resettled in Chongqing and Xu's company will be responsible for their resettlement. That is why there is such a estate fever. One *mou* of land cost ¥30,000 in Jiangbei last year: now it costs ¥200,000. I can't convert these figures to Hong Kong-style profits in my head, but he has gotten my attention.

Toronto has real estate developers with little doing these recessionary days. Can foreigners buy real estate? I ask. The answer is in most cases they sign a 70-year lease. Our friend volunteers that perhaps it is a bit late to invest, prices are already too high.

Mr. Xu has a project: five mou of land in Jiangbei, partly occupied by a rudimentary furniture factory. A small investment, say US$150,000, can make a quick profit. A joint venture could import a painting line from Italy (we suggest Canada? a used one?). The real estate boom has triggered the demand for furniture. The company has skilled workers, it would be labour intensive (good reason for quick approval from city officials). The kicker is the 18-floor office building that could be built above the factory. Office space earns ¥200 per square metre. Mr. Xu will manage the factory, the approvals, everything.

Why, with ¥100 million in revenue, does he need us? He needs a foreign partner to avoid taxes, to go abroad . . . . We have the feeling that

Xu's thinking is as old-fashioned as Mr. Deng's before his conversion in Shenzhen. Chinese businessmen have no trouble coming to Canada. As Zhongling is fond of saying, "Money is enough."

(Postscript, October, 1993) The boom is bust. Dust blows from the immense basements hacked from the limestone. The cranes are still. The Bank of China lowered the boom on an overheated economy and none too soon but this is the free market system. Just wait!

Income: White, Grey and Black

(Robert, Chongqing, September 15, 1992) For me this whole thing began when Caroline casually mentioned that she wanted to talk about "this China thing." A couple of weeks later we got together for lunch in a Toronto Queen Street dive and Caroline, in her normal manner, beginning in the middle, told me about this plan to build a huge dam on the Yangzi River near where she and Norman had been teaching in China. She mentioned the idea of doing a book. I thought she wanted some feedback since I had worked on a number of heritage studies, but then she said, "So, do you want to go?"

I'm quite sure I spilled my beer at that moment. "It is not completely arranged yet," she said, "but you should get your shots and book a flight and that sort of thing."

At this point in my life I have very little wanderlust. But China was the one place I really did have a yen to visit. I had taken a course in Chinese history as an undergraduate over twenty years ago. It was taught by two remarkable young women at the University of Western Ontario and they had been powerful influences on my personal growth. Books like William Hinton's *Fanshen* and the poetry of Li Bai and Du Fu left permanent grooves in my mind. In his story of the Communist Revolution in a rural village, I remembered Hinton's description of the stain on a peasant's thumbnail from crushing lice. Today, less than three months after Caroline's first mention of this project, I am standing on a road in Sichuan looking at people waiting for a bus, and seeing those marks of the hard life that so many Chinese peasants still live.

In the first week of August, 1992, Caroline called to say that the trip was all set. At the end of August I defended my Master's thesis in Urban and Regional Planning and a week later flew to Hong Kong. My introduction to the Orient was made with the help of two Canadians, Bill and Sheila Purvis. Bill is another NC Press author whose book *Barefoot in the Boardroom* is a delightful account of his "adventures and misadventures" managing a joint-venture manufacturing company in mainland China. Sheila works for the World Health Organization and is administering a teaching programme for doctors in China. Between them they have a wealth of knowledge and experience, as well as a genuine admiration and affection for the Chinese people. When they put me on the plane for Chongqing I was as well-prepared for my own adventure as I could have been under the circumstances.

Nothing, however, can really prepare one for China. It was still light when I took off from Hong Kong and the sky was clear. From the plane window I saw the urban expanse of Guangzhou (Canton), the broad West River, the patterns of fields and smoky smudges that marked the sites of factories or coal-fired generating stations. Then in the last rays of the sun I saw the spines of the mountains of Guizhou Province. From this bird's-eye view I could see how much of this country was rugged upland, and how little was fertile plain.

I suppose I was prepared for the worst after the stories of teeming masses, unsanitary streets and crowded living conditions. Suddenly, the plane was on the ground — I had seen no lights of a city as we approached and indeed the airport itself was in virtual darkness. I hadn't realized until I saw this just what a shortage of electrical power really meant.

Caroline was waiting for me with a car, and we set off immediately for the hotel. The first sight I had of the city was a surprise. Apparently we were driving through a "new economic zone." This was not what I had expected. The streets were deserted, but in the absence of much light the place didn't look bad at all. The Chongqing Guesthouse, when we arrived, turned out to be a complex of buildings whose Chinese-style roofs were outlined by strings of coloured lights. The effect was delightful.

We were met by the man who was to be my almost constant companion, translator, host and friend. Huang Zhongling is the son of one of the three Chinese men who had lived with Norman's family for six years, when the Endicotts were missionaries in Chongqing in the 1930s. Zhongling was considered a nephew in a sense that holds considerable gravity in China, where family relationships are paramount. Zhongling is about my age and, although I found him gracious, well-mannered and quite accomplished in his spoken English, I think he was initially quite disappointed with me. He is tall, handsome and in trim physical condition. He was also impeccably dressed in a neat coloured shirt, pressed trousers and polished shoes. I, on the other hand, am pretty short, for a foreigner, overweight by Chinese standards and, with a beard and drab-coloured, flight-wrinkled clothing, I was not the picture of Western high style.

While the Chongqing Guesthouse has a Chinese-style exterior, the inside has been renovated. It looks like a Holiday Inn. It was the last such comfort I would experience in China. As soon as I was in my room an attendant appeared with a large thermos of boiled water for tea. The following weeks would teach me the importance and many uses of the water supplied almost everywhere in that fashion. I turned on the TV. On one station there was nothing but advertisements. These seemed to be mostly Hong

Kong-produced inducements to buy a vast array of cosmetics, electrical appliances and other consumer products. The next station featured a game show from Beijing. The local channel had a pleasant-looking man talking very seriously. Zhongling explained later that this guy has a nightly, prime-time show about investing in the stock market. The last station was CNN in English.

From my window I could see the rest of the hotel precinct. On the far side of a large courtyard where many cars were parked was "The Recreation Centre." The neon sign proclaimed the title in English as well as Chinese. This turned out to be, among other things, a collection of upscale nightclubs. The waitresses in the "Texas Bar" were dressed as cowgirls, and the cover charge was equal to a week's wages for a Chinese office worker. Zhongling had never been to the place before.

"How can people afford to come here?" I wondered. "I've been told that Chinese workers only make ¥250 to ¥500 a month (US$50 to $100) and it costs ¥25 just to get in here?"

"I need to tell you something about Chinese characteristics," Zhongling began. He was to start many subsequent explanations in that way. "In China there are three kinds of income: White income, or wages; grey income, which is something else; and black income . . . . I think you might imagine what is meant by black income."

He said that he didn't have any of the darker shades of revenue to be able to afford The Recreation Centre. I might have been able to afford some recreation, but it is a good thing that I didn't. Just before I went to bed I looked down at the courtyard and saw some attractive young women. One of them saw me and waved. A few minutes later the phone rang. A pleasant voice asked me a question in Chinese. My stammering brought a giggle and the line went dead. The question was a pretty obvious one in just about any language. I found out later that had I responded positively a young woman might have appeared at my room, followed quickly by the police. This was not the country I had read about in *Fanshen*.

Following my one night in the luxury and delicious danger of the Guesthouse I moved to Zhongling's house. This was a very generous offer on his part and represented a considerable saving for me. At first, however, I had no idea of the significance of this gesture. He told me with considerable satisfaction that, as far as he knew, no foreigner had ever stayed in a Chinese house in the city before. "Ten years ago," he said, "no one would ever have thought this could ever happen in China."

Zhongling's three-room flat is spartan by Western standards, although he has a refrigerator, a fan, an automatic washer and a TV. What he does not have is a shower or a toilet. Baths are taken using a washcloth standing in front of the sink. This Zhongling did two or three times a day. The call of nature is answered down the street in a public lavatory at the cost of a few pennies. The couch I slept on is the most spartan element of all. It is covered with a kind of bamboo curtain; quite hard.

With the help of an often-referred-to Chinese/English dictionary, my conversations with Zhongling are growing in complexity and depth. While I have abandoned any attempts at speaking Chinese beyond *pijiu* (beer) and *xiexie* (thanks), Zhongling's confidence and delight in speaking English grows daily. We are now able to joke and carry on freely. Tomorrow we leave on our odyssey down the Yangzi. Today we visited the tourist office to buy our boat tickets. There was a young woman dressed in a bright velvet dress. I have learned that Zhongling is divorced and we have begun to exchange some comments on attractive women. I pointed out the young lady in question and asked if he was at all interested in someone like that.

"You know what we say about that kind of dress?" he asked.

"No, what?"

"It's good for polishing your shoes."

I guess he didn't find her appealing.

The Canadian Connection

(Robert, Chongqing, September 16, 1992) In 1989 the Canadian government accepted the report of the Canada Yangtze Joint Venture (CYJV) group, the *Three Gorges Water Control Project Feasibility Study*, prepared by a consortium of consulting companies and public utilities, including SNC/Lavalin of Montreal, Hydro-Québec and BC Hydro. The Canadian International Development Agency (CIDA) had commissioned the $14 million study, but the document was only made public when an environmental group applied under the Official Secrets Act. The Canadian government praised the work as "a world-class effort which would help the Chinese government make an informed decision on whether or not to build the dam." It was hoped, of course, that Canadian firms and utilities might benefit from future contracts should the dam proceed.

Caroline, Ruth and I made ourselves familiar with the available documents before this trip. We did not have depth of expertise in engineering, wildlife management, economics, flood control, environmental protection and most of the other disciplines involved in the CYJV study. It is not our intent, however, to take part in either a protest against the dam or an endorsement of it. The task we set ourselves was to bring our own understanding and sensitivities about culture to the area, and to learn as much as we could about what forces are going to affect and change the Three Gorges region. What can usefully be done by concerned people, both inside and outside China, to mitigate the cultural impacts? We hope to approach these questions, and the region, with objectivity and a genuine concern to learn, and then to share the excitement and discovery of our journey.

The most important component of the CYJV report is Volume 8, Appendix G, the cultural site survey. We used it, in part, as our guide. But we were much more interested in what the local people regard as important.

As we get ready to leave Chongqing I am particularly well rested, having enjoyed the best sleep since my arrival in China four nights ago. Zhongling came in last night after visiting some of his friends and took the hard bamboo curtain off the couch. He must have thought previously that I wanted it there while I, in my turn, had thought that was how it was done in China. Each of us was too polite to question the other on the subject. In this regard, Chinese and Canadians are much alike. We would rather be uncomfortable than make fools of ourselves.

I have noticed along the riverbank where we will be catching the boat today, elevation marks in metres painted on the stone. In the protracted discussion about the building of the Three Gorges Dam, one of the main topics of both engineering efficiency and potential environmental effects has been the depth of the planned reservoir. The engineering studies refer to the NPL, or Normal Pond Level, as the central parameter defining various proposals for the project. Competing interests are expressed through different concepts of what the optimum NPL should be.

The deeper the reservoir, the more hydroelectric power can be produced. So the Chinese power authorities would like to see a reservoir whose surface would be normally maintained at 185m or more above sea level. In fact, that is the level on which their most optimistic projections of power output are based. The flood control authorities favour a much lower normal level, perhaps 150m above sea level. That would allow them to impound even the maximum flow of the river during the flood season, which occurs when winter snow accumulation in the Himalayas melts during summer. The reservoir level would of course rise dramatically during that runoff, but it would have to be kept low prior to the flood in order to absorb the excess. This lower NPL, however, would greatly reduce the potential power output.

When it comes to river navigation, there are at least a couple of considerations that relate to the depth of the reservoir and the regulation of water flow. Shipping authorities would favour minimal seasonal variations in water levels. They are also concerned about siltation, especially at the upper end of the reservoir. There the fast-flowing, silt-laden waters of the upper river will meet the relatively still waters of the lake formed by the dam. The tonnes of particulate matter carried by the river will settle out, and continual dredging will be required to keep ports such as Chongqing operational. All of these arguments, of course, avoid completely any discussion of environmental or cultural impacts.

So the debate about Normal Pond Level has dragged on. The Canadian-sponsored feasibility study recommended a NPL of 165m. This, Canadian engineers felt, was the optimal compromise between the various interests. After the go-ahead for the project was secured at a meeting of the National People's Congress in April of 1992, however, Premier Li Peng ordered preparations for the dam to begin with an announced NPL of 175m above sea level. It appeared that the power supply lobby had essentially

taken the sweepstakes. So as we explore the steps descending to the Yangzi
river boat piers in Chongqing, we read the elevation markers painted on the
ancient stones and try to fathom what this all means. As nearly as I could
tell, the river level at Chongqing in September 1992 stood at about 140m
above sea level. The 175m mark was quite close to the top of the stairs.

Now what I see in Chongqing, and will see over and over in every
town, village and coal tip along the river, is this: between the river level and
the first permanent structures, built higher up on the steep banks, there is
a flood zone. This is an area with no fixed structures except stone stairways
and ramps. I am told that when the river rises in the summer, as it always
does, the water swallows these walkways and laps at the foundations of the
lower buildings. In bad years when serious flooding occurs, the narrow
winding streets closest to the river are inundated and any number of the
houses in these quarters are swept away. A few days ago we visited one of
these old riverbank neighbourhoods, not on the Yangzi but on the Jialing,
the major tributary that joins the big river at Chongqing. I saw a commem-
orative marker on a building showing the high-water mark of the 1981
flood. I look at the size of the adjacent valley and I find it hard to comprehend the volume of water that it would take to fill it to this depth.

What is being proposed, then, is to permanently fill this valley to near
the level that it reaches in seasonal flooding, although that level will be
allowed to run down prior to the annual surge of Himalayan melt-waters.
This will mean that the traditional relationship of the city to the river will
be drastically altered and, in the years when catastrophic flooding occurs, it
will probably be worse than ever in Chongqing.

A Once and Only Chance

(Kailin, Chongqing, October, 30, 1992) Caroline and Robert were very interested in finding out what the Chinese government intended to do about
the many archaeological and cultural sites we would visit on our trip. Back
in Chongqing, after they had returned to Canada, I contacted government
officials to look for the answers to their inquiries.

Since April 3, 1992, when the National People's Congress formally
passed the resolution on the Three Gorges Water Conservancy Project, the
National Bureau of Cultural Relics has been planning with local professionals in Sichuan and Hubei, and investigating historic sites below 177m elevation in the reservoir storage area. This area will be 600km long and will
touch on 22 counties. At present, the survey has recorded 842 points of
cultural interest: 376 places with relics above ground, and 466 with relics
underground. No development project so far has dealt with so great a number and such a large variety of relics.

The Three Gorges cultural legacy project will last a long time and spend an unprecedented amount of money. However, it is a salvage operation and a once and only chance. Although the project has been undertaken wholeheartedly, current methods and resources restrict what can be done. According to the Three Gorges Relics Protection Project syllabus, only about ten percent of the general area, 23 million square metres, and thirty percent of the ancient burial sites will be studied. The most important ancient cultural sites will be excavated, the less important ones partially excavated.

To fully document cultural relics, books will be published and new museums will be built. A Three Gorges Historical Museum in Wanxian will have the status of a provincial museum. The new town sites will maintain their historical style and settings. Such relics as the Big Buddha at Danzisi in Chongqing, the City Wall in Fuling, Shibaozhai Temple in Zhongxian County, Zhong Luo Temple in Wanxian, Yaokui Ta Tower in Fengjie, and Baidi City will be left on site, reinforced and restored.

The sites to be relocated will be moved on the principle of "no change in the original shape." These include the Zhang Fei Temple, the Qu Yuan Ci (Ancestral Temple), the Qiufeng Ting (Autumn Wind Pavilion), and the Da Chang Ming Ju (historic house). Some structures which are not intact, but which contain authentic parts, will be partially moved: the City Wall at Yunyang, Yuwang Palace at Shizhu, the Hujia Big House at Zigui and Wuyitang Historic House in Xinshan County, are other examples.

The Mighty Dam

(Ruth, Yichang, October 29, 1992) At Yichang, I saw one of several models of the new, controversial, multi-purpose, 185-metre high dam to be built on Zhongbaodao Island upstream. It would be nearly two thousand metres wide. Two power houses would produce 17.68 million kwh with an annual production of 84 billion kwh. The navigation channel would have five steps, each step lifting or lowering ships up 22 to 25m, a total of 100m when the dam was finished. The cost was estimated at US$10 billion.

The new dam would create about 12 times the electricity churned out by Niagara Falls, Canada. Foreign tenders have now been called. Actual construction started the summer of 1993, and it should take about 18 years to build (we keep hearing different estimated completion dates). I thought of the power cuts, the brownouts and the days of work lost because factories could not operate. I thought of the need for jobs, not only for Beijing's 2 million floaters, but the 27 million people barely surviving on less than ¥200 a year. Roughly US$35, this is China's poverty line. When one talks about China, population figures loom huge; incomes sound low. Thank goodness China has one of the best population control programs in the world. Per capita income could have been worse.

The new dam would cause the flooding of 140 towns and 13 county seats. Of this, 23 800 hectares of farmland would be lost; 1.13 million people would have to be moved. Fu Xiutang, Vice-chairman of the Yangzi River Water Conservancy Committee, has been quoted as saying that 10 million people have already been relocated in China since 1949 because of similar projects. And the *New York Times* mentioned the floods of 1991 killing 3,000 people and leaving 3.2 million homeless.

Supporters of the dam are saying that the movement of the 1.13 million would save greater human displacements because of future floods. It is a good argument, except that the previous 3.2 million were from all over China and, after the dam is built, who would ever know if it is true? But there is no question that destructive floods have been endemic in China.

Opponents of the dam say the migration forced by the new dam is morally wrong, a disruption of ancient lifestyles. Are they trying to keep Chinese people in the Stone Age? There have been many migrations in history, both voluntary and involuntary. My own father and grandfather were driven out of South China by poverty in the early 1900s. Fortunately for me, they were able to make a new life in Canada. Should the poor of China all migrate to Canada? Migrants all over the world have adapted and survived and many have prospered. Have the critics really talked with the people to be moved?

Critics also argue that electric power should be conserved instead. While it is true that much electricity is wasted, many parts of China are suffering blackouts and brownouts. There is not much electricity to conserve. Opponents also cite the buildup of silt in the reservoir area, but officials say they are planning to flush the silt through the dam in summer. Critics also talk about the vulnerability of one target that could be destroyed in wartime. Perhaps a series of smaller dams, rather than one giant monument to the regime, would be a better solution. . . .

Rebel Emperors and Buddhist Nuns

(Caroline, Chongqing, September 16, 1992) It is past 6:30 and the men have not arrived. Zhongling lives just around the corner, but it is a corner where five streets swirl together. I pick the right one, but then I realize that, although I have been to visit his Aunt Alice on the main floor half a dozen times, I have no idea where Zhongling's apartment is. The Huangs live in remnants of one of their old family houses, a ramshackle building of stone, half-timber, wattle and concrete, squeezed between newer buildings. Zhongling's 85-year-old grandmother had once shown me the bedroom there that had become her prison during the Cultural Revolution. She had been briefly famous for making speeches on women's rights after Liberation, but when the campaign against "capitalists" (not to mention those

who knew foreigners) swept Chongqing, even she was a target. There is a family story that the whole apartment was only returned to the Huang family when Norman's father, the famous teacher Wen Yuzhang, insisted on visiting them after the Cultural Revolution.

I meet Robert and Zhongling coming down the outside stone stairs. Norman has charmed the harassed waitresses. This will be our last "American" breakfast for some time. The eggs are swimming in fat, sunny side up, the whites not cooked. The coffee is weak. Never mind. From here on, breakfast will most often be my jar of Nescafe, and what we can scrounge.

Zhongling insists on departing in style in a big ¥ 12 ($2)cab. Our decisions as to when to splurge and when to save money will reveal some interesting cultural differences between us. The streets are dark and almost empty until we arrive at the dock area. Chongqing's passenger terminal is only a flamboyant architect's drawing. Like so many projects in Deng's New China, it awaits a Japanese or Taiwanese financier with deep pockets. But perhaps now building the facility will become a matter of civic pride: Chongqing is the western terminus of the commercial passenger run to Shanghai.

The road above the steps leading to the dock is already jammed with vehicular and pedestrian traffic and the bus and taxi drivers lean on their horns. People are carrying their suitcases and bundles, workers are arriving for work, peddlers are selling ripe green and red apples done up in plastic mesh bags. Everywhere people are grabbing breakfast on the run, squatting on their haunches, eating noodles or *baozi*, staking out a place to wait.

Here we sit, until Kailin finds us. Bad news: the boat will be delayed until 11:00. We might get on at nine or ten. As expedition leader, I take command. There is no place to sit except for the side of the road; it is drizzling; there is no bathroom. We will ask the taxi to turn around in the traffic, and the women (Li has found us) will return to the hotel. Zhongling and Robert will follow in a small cab.

Norman is having his after-breakfast nap. I clear off the second single bed, double up with him, and motion to the other two to lie down. They have had much less sleep than I (Kailin has come all the way from the suburbs by bus) but they do not want to rest. Li makes phone calls from our hotel phone (at least the calling-out part works) and the irrepressible Kailin cannot stop talking. For me, getting older means calculating the shortest distance between two points and taking advantage of every opportunity to rest! The men return at 8:50. They have been out rounding up small change. We will have problems buying apples up country with ¥ 100 bills.

This time Robert and I avoid the lineup of cabs. There is an almost empty *kangfulai* (minibus) right at the curb. Chongqing has a two-class urban transportation system (three classes when you count the company cars). The public buses defy description. The *kangfulai* system is run by private enterprise, charges about five times the public rate, but guarantees a seat. We pile gaily in, and less than ¥1 each gets us about as far as the cab would have been able to go.

The boat for our pier has still not arrived. I had been pushing for staying longer in the hotel: we have tickets, don't we? We do, but Kailin is determined we get a good cabin. So, just wait. I head for an unclaimed stone block sticking out of the sandy bank. "No," Zhongling says, "it has been used as a toilet." The best we can do is a few feet of sidewalk. Out comes *China Daily*. We spread out a page for the mound of luggage and we each take a corner to sit on. Some of our fellow travellers have been camped out here like this since 7:00 a.m., playing cards and picnicking.

The drizzle, the tail end of the monsoon that a day or so ago killed a thousand people in Pakistan and India, has almost stopped. "Semi-tropical" Chongqing is on the northern edge of the monsoon zone, which means it gets plenty of gentle rain, but, rarely, any wind. It is September, and the high water time has just passed. At low water in late spring we would have had to pick our way across a tremendous expanse of gravel before boarding the ferries.

To our left lie the famous stone steps of Chaotianmen, made famous by Sommerset Maugham. Here, over untold centuries, porters have toiled goods from river boats up to the city, and honey buckets of human waste down to the small sampans, for fields accessible only by river. One patient footfall after another has worn smooth the filthy, muddy stairs. Eyes, sometimes bright, sometimes blank; iron calves with muscles like ropes; the flimsiest of footwear; ragged blue cotton jackets and pants; back muscles straining under thick bamboo yokes slung with bamboo baskets or balanced ropefuls of boxes. Large cargo is not offloaded here any more but, still, the labour of peasants, many thousands of illegal immigrants from the countryside, is cheaper than machines for some things in Chongqing.

Chongqing is proudly called "Mountain City." There is little flat land here, really none at all, even on the famous goose neck between the Yangzi and the Jialing rivers. The steep sides of their valleys continue on up into low mountain ranges. Roads snake along these mountainsides, one above another, with very few cross streets and traffic lights. Once the traffic jams, and it frequently does, there is no escape. Caves have been dug into these cliffs since time immemorial, for their coolness in summer and shelter in winter, for protection against Japanese bombs during the war and, today, as a solution to the shortage of land. Dig a cave and you have a storefront on a main street. Parking is a problem, but then cars have chauffeurs here.

Chongqing first appears in historical records as a settlement of the Xia Dynasty, in 2200 BC. It is mentioned about 1100 BC as being part of a feudal kingdom and, in 375 BC, as in the "Kingdom of Pa"(*Ba*). Jianzhou, its capital, was located fifty kilometres away. The first wall was built in 320 BC. Ba was the name of the area in Han times. The name Chongqing (meaning repeated or great good luck) dates from AD 1188. In 1370 the wall was rebuilt of solid stone, eight kilometres in circumference, and breached by 17 gates.

Chongqing stands strategically at the west end of the Gorges, guarding the entrance to the fabulously wealthy Sichuan Plain. It was forcibly opened to the outside world by the Treaty of Yantai in 1890, and was considered to be within the British sphere of influence. The Europeans had nothing to gain by trying to rule the vast and unruly land of China, but much profit was made by controlling its trade. We have a picture in our dining room of British gunboats in front of Chaotianmen. An American consular report for 1926, the year Norman was born, notes that Chongqing's population was 500,000 including about two hundred Europeans. France, Germany, Great Britain, Italy and the United States had opened consulates. The Japanese did not develop their concession but, as today, they did the most business because they paid the most attention to ascertaining the needs of the local market.

At the water line, in the early decades of the century, stood an office of Maritime Customs and the "godowns" (warehouses) of Standard Oil and other foreign companies. The British and American naval canteens lay in the southern suburbs, as did the pleasure gardens of the guilds, the embassies, and the residences for the legendary Hong Kong hongs of Butterfield and Swire and Jardine Matheson. These southern suburbs were cut off from downtown by the treacherous Yangzi, and accessible only by ferry, but they offered pine-covered mountains and respite from the hubbub of the city. After the Nationalists moved their government to Chongqing in 1938, Chiang Kaishek and his American allies, generals George Marshall and "Vinegar" Stilwell, built their villas here.

"In ancient times," a wall ran all around the old city. The many gates are gone now and so is the wall, except for remnants that have become part of the river retaining walls. Below these masonry walls cling ramshackle houses of bamboo poles and matting, their floors propped up and out with stout bamboo. Each summer when the water rises they must be dismantled. Like so many things in China, they appear to be centuries old. New buildings, however, are being built at a furious pace, on the mountainsides and on the river side of the roads, wherever a terrace can be hacked out or extended with concrete. Many will be in increased jeopardy of flooding from the reservoir.

The fear of flood is real. The two very serious floods since Liberation linger in the memories of my Chongqing colleagues. A small park at Chaotianmen has a plaque commemorating the high water level of July 16, 1981. Robert and I finally found it yesterday, but only at dusk. It should have been simple to get there: head down any stairway toward the "lower ring road" and bear left. But there was no direct way through the crazy-quilt of stairways and passageways. We almost blundered into a mourning family's storefront livingroom before realizing that the paper wreaths out front were not just advertising for one more funeral supply shop. The old woman's body was laid out only a metre from a narrow passageway that for hundreds of people is the road in front of their home.

It is impossible to imagine this immense valley filling up in less than 24 hours, and overflowing these banks. We had fled the heat of Chongqing's summer in 1990, but 1990 was not a particularly high water year. But in 1981, Chongqing suffered its worst flood this century, a "hundred year flood." Heavy rain for six days caused an immense increase in the rate and speed of flow. The water reached a height of 194m: 28 died, 137,000 people were affected and damage was estimated then at $5.4 million. Early warning gave most people time to evacuate. Armies of volunteers shoveled away the huge quantity of silt deposited in the valley.

Flood control is the most controversial but compelling reason for building the dam. Electricity is needed and its benefits are easily measured. It could be made other ways. How much protection can reasonably be provided against a flood that may occur once in a hundred or even a thousand years? A flood is an incredible force of nature and the water has to go somewhere. Either it will be held behind the dam, to flood Chongqing and other cities on the reservoir, or it will be let out to flood cities downstream.

Theory is fine, but some hydrologists believe that flooding at Chongqing will be made worse by the dam. The current "natural" water level is 158m. The lowest Normal Pool Level (NPL) anyone is suggesting is 160m and 185 is more likely. There is agreement that silting will form a delta at the end of the reservoir, raising the five-year flood level here by about four metres. Keeping the reservoir at a minimum of 160m, except for flood time, would eliminate all but one of 32 bottlenecks for shipping and all 12 winching systems. However, at the flood-time level of 140m, there will still be five to eight bottlenecks. Is this enough to make shipping as reliable and cheap as Chongqing's city fathers hope?

It is well after ten o'clock when the boat arrives, unloads and eagerly we make our way down the stairs. Kailin and Zhongling work their magic and get a satisfactory cabin: quite far forward, bunks near the door and close to the fresh air, for there are no non-smoking cabins. Our third-class tickets have been carefully considered: they are cheaper than second class (which is first class), and more than tolerably comfortable. Tickets for lower classes are not available from the tourist agencies.

We foreigners are blissfully unaware of the class realities on Chinese ships. It will soon, however, become clear that these iron bunks, in cabins which run the width of the ship, are the only places to sit, except for the passageways and stairways, the very small aft decks, and the brief and noisy hell of the dining room. Getting on and off the boat requires picking our way over the bodies and bundles of those who can afford only fifth class. In fact, most of our trips are short, so it doesn't much matter what class we have.

Soon we are comfortably stowed and under way and ready to say good-bye to Chongqing. The boat pushes out and swings right around. Ahead is the place where the green waters of the Jialing River meet and mix with the brown, silt-laden flow of the Yangzi. The Jialing is one of the biggest and most important of the Yangzi's many tributaries.

Now, on the east bank (for here the river temporarily flows north), are the gaily-painted porches of the Ciyun Monastery, begun in the Tang Dynasty as a temple and hostel for pilgrims to China's famous mountains. Buddhism reached China from India in the early flowering of the Han Dynasty, but it was during the fourth to sixth centuries, China's own dark ages, that Mahayana, the mystical school of Buddhism, took root here. Monasticism was inimical to the Chinese tradition of filial piety and parenthood. Buddhism's subtle and metaphysical doctrines of enlightenment and salvation by faith did not take root in the essentially practical and materialistic Chinese intellectuals (who saw them as an alien but interesting novelty), but in the religion of the common folk.

Two black panting lion-dogs guard the river door to the monastery. Drum towers and walls, as impressive as any fortress, cling to the rocky slope, commanding spectacular vistas of the two rivers. The main hall houses one of the largest jade Buddhas in China (1.87m high, 1 500kg), believed to have been brought from Burma by a layman, Xiang Qiangfa. Precious relics include a rare 6,000-volume edition of the *Jiripitka*, printed in the Song Dynasty. The hall is hung with brilliant red prayer banners: gifts from the pilgrims of today, from Taiwan.

A white jade statue of Sakyamuni as a naked child stands stiff and small in a fountain, inviting a coin, promising a boy-baby. The tree behind him is said to have grown from a cutting from the same Bodhi Tree that sheltered the Buddha-to-be from fire and darkness when the evil Mara tried to vanquish him on the night of the Full Moon of May. Beyond, in the garden, are the cream-coloured conical stupas of two monks said to have brought sacred sutras all the way to Chongqing from India. At the highest level, a shrine cut into the rock once sheltered monks laid out to await their passage into the higher stages of consciousness. When their souls had finally left their bodies, they were cremated there. This has not been allowed in recent times, although Buddhism is undergoing a minor revival in China today.

Ciyun is one of the few monasteries in China with both monks and nuns, although all of the nuns I saw were very old. Their vegetarian cooking is world famous, they say, and they are now marketing their monastery as a retreat to tour groups of pilgrims from Japan and Taiwan. Paradoxically to us, vegetarian cuisine in China is judged by the cook's skill in duplicating the taste and texture of meat dishes with tofu and other vegetable products! Ciyun is not listed among endangered monuments in our report and Li is quite firm that it will not be affected. Its bottom porches and passageways must have been flooded and rebuilt many times over the years, but the Great Hall of Buddha will be safe.

Above Ciyun Monastery, commemorated with a street sign, is Huang Family Lane. It is not so much a street as a passageway, paved with stone slabs. Here Zhongling's family, once one of the wealthiest in Chongqing, kept a traditional landlord's country home. Zhongling's great-grandparents had lived there. His great-grandfather was the ruling member of the family but by the time Norman remembers visiting there in 1936 or 1937 with Zhongling's father, Huang Gongwei or "George," the old man was an addict, smoking opium all day. As the boys entered via the gatehouse, they would pass his huge coffin, cut from a giant tree with the outer sides of the planks left round, placed there, as was the custom, in anticipation of his death. Norman remembers that George's father treated his family to a sumptuous dinner there.

But George's father was a thoroughly cosmopolitan man who had visited Europe and America and wanted himself, and his sons, to learn English. Starting with large landholdings, they had diversified into banking and coal mines and steamships. Zhongling's grandmother came from a wealthy banking family, the Yangs. (Today Zhongling can point out the buildings or the places that were the headquarters of the family firms.) George and two other boys had come to live and go to school in Norman's family house and, ever since, the Endicott children and their boarders (and their siblings) have called themselves brothers and sisters, and Norman's father "Uncle Jim." And so, Zhongling came to our apartment every Thursday night in 1990 to visit his "aunt" and "uncle."

Beyond Huang Family Lane, farther still up the mountain at the little hamlet of Huangginmiao, are the missionary residences and the Canadian Mission Hospital which was built later. The hospital still flourishes. The houses, where Canadian missionary families lived in relative comfort and some style, have been stripped of their elaborate turned woodwork, and their vast porches have been bricked in. Long ago, they were carved into living quarters for hospital staff. Norman lived in one of those houses.

A bit further along, and high out of flood's way, the Marine Francais still stands. Li, Kailin, Francoise, Zhongling and I had visited it in 1990. A stone plaque in the courtyard commemorates the French architect who built it in 1902. Its Mediterranean arches, verandas and French doors were designed to catch the breeze: only the entrance gate is Chinese. It was first an office and residence for French naval officers and then the home of the French consulate. Today it is a factory office, and in very bad repair. The dining hall, its elegant arching windows broken, is used for storage. Behind it, a warren of dark buildings encloses a wooden courtyard used as a stage in the Cultural Revolution. Very few foreign buildings remain, or indeed, any buildings from before WWII. The Japanese bombing was devastating.

The skyline of downtown Chongqing is behind us now, flaunting its flamboyant, curved and brightly-tiled modern towers, in what I call the "Joan Miro style." How different it is from the view celebrated in the 1890s by the intrepid businessman and steamship visionary, Archibald Little: "rock, river, wood, and temple, crenelated battlements, and uplifted roofs . . . altogether one of the most perfect pictures of river and mountain scenery, enlivened by human activity, that the world affords" (Little 1898:164).

Sooner than I would have thought (it took a couple of hours to reach it by van) we came to the riverside Big Buddhas *(Da Fu)*, in the suburb of Danzisi. The bottom Buddha, from the Yuan Dynasty (1357-65), is listed as at 179m above sea level and the second one, from the Ming (1403-24), at 190m. It is hard, intuitively, to believe that Chongqing, so far from the sea and in mountains, has an elevation so low.

The Yuan Dynasty Buddha was actually built by an upstart local rebel emperor, Ming Yuzhen. His false tomb (a decoy to fool grave robbers) was recently found in the compound of a textile factory, north of the Jialing. Ming was born of a peasant family in Hubei. Even as a child he burned with desire to overthrow the rule of the cruel Mongols of the Yuan Dynasty, the heirs of Genghis Khan. In 1353 he successfully led a local uprising and by 1357 he had proven himself as a warlord.

The ambitious young man then learned that the Yuan officials in Chongqing were weak and divided. Ming took his army by boat, through the Gorges, and appeared suddenly before Chiaotianmen. The imperial army fled. By 1358 he had captured Leshan and Chengdu and the peasant insurgents found themselves in control of a kingdom in Sichuan. In 1360 Ming was chosen emperor and Daxia State was established. In 1363, the capture of Zhongqing (now Kunming) added a huge area to his realm.

Ming was welcomed as a liberator. His army was admired for its good discipline and his administration was a model of austerity. He confiscated land from the landlords and the Yuan herders and redistributed it, reduced taxes, abolished the inflated Yuan currency, and stabilized prices. He encouraged education and the religion of the Maitreya, the "laughing Buddha," and he gave money so *Da Fu* could be built. Ming Yuzhen died, some say he was assassinated, at age 38 in 1366. His son was only ten and by 1371, the new Ming emperor, Hongwu, was able to lead an army through the Three Gorges and up to Chongqing. Daxia State collapsed. Hongwu built his Buddha above Ming's and there together they remain.

I had seen a faded picture of the Buddhas in the Chongqing Museum. We visited them in 1990, as an afterthought. Norman had taken the foreign teachers on a nostalgic visit to the Canadian missionary sites. The director of what had formerly been the mission hospital in Danzisi offered us oranges and tea. One of the doctors volunteered to take us to the Buddha: patients and their relatives go there to pray. The shrines and their derelict gardens are now within a factory wall. Chiang Kaishek had rebuilt the pavilions during the war but now all but the places of worship are sunk in decay.

Emperor Hongwu's Buddha and disciples are housed in an elaborate curved-roof pavilion, and swathed in lengths of red cloth. Offerings of fruit and oil and incense deck the altar. Beside the ancient Buddhas are new plaster, roughly hand-painted "local gods" offering good luck and, again, sons. Ming's Buddha sits alone, powerful and defiant, his arms braced against the flood. Riverside Buddhas, sometimes of gargantuan size as at Leshan and Binglingsi, were built to tame the old gods of wild and unruly rivers.

The excitement of departure has subsided. For over an hour we will still be leaving Chongqing, which has a huge municipal area with satellite towns larger than most Canadian cities. There is time now for the first meeting of our little expedition, out of the earshot of bosses and others. Robert has prepared a written explanation and plan for our project so that our friends will know why the mad foreigners are taking pictures of people who are not their relatives or friends, and of places that are not famous scenic spots. We are interested in culture very broadly defined, including the way people live, handicrafts, local industries, products, and traditions. We are aware that we will have to go inland, away from the river, in order to see very much.

Our friends think this is very nice, but their concerns are elsewhere. Zhongling points out that we must be very careful: some places we are going are dangerous for foreigners (actually for foreigner's possessions). The people are very poor and there will be no police to protect us. We are warned that "the common people," one of Kailin's favourite phrases, are very poor here and very primitive. (Just to keep things in proportion, we scarcely see a person in uniform and nothing is stolen.)

Li says look a lot, say little. We are tourists, on a little holiday. There must be no talk of an "expedition" or a "project." That is enough business. In fact, neither Kailin nor Zhongling want to be tied down all the time as translators and Li is very clear that her role is archaeological expert and negotiator of our *guanxi*: connections. Kailin will be the source of constant good cheer, no slouch with *guanxi*, and our chief expediter, our charmer of boat captains. Zhongling will be a negotiator, especially when a masculine face is an advantage, and fill in the copious forms required to register at the most modest hotel. He will be a good and hard-working translator.

The cabin has ten iron bunks, furnished with mattresses with bamboo mats on top of the sheets for coolness, a big terry towel and a pillow, all neat and tolerably clean. We have few cabinmates. There is a desk, where we can pile our plates of food and tea, a chair, and a sink. To my amazement our friends wash their dishes in the tap water! A thermos holds hot water so we do not even have to go out to the boiler in the passageway for tea.

The loudspeaker system plays Viennese waltzes. It is lunchtime and all but me go to the dining hall to fetch lunch back to our bunks: I guard our worldly possessions. Our Chinese friends have their own containers, cups and chopsticks. Robert and I have to use the pernicious foam takeout containers which, like almost everything else, are eventually tossed into the river. The dining hall is in the stern. It is completely of metal, including the tables and chairs. Mealtime is short. You buy tickets, line up, thrust your bowl in through the crowd for a scoop of rice from a metal drum, and a small ladle of vegetable and meat on top. The dining room is packed (we assume largely by those who have no bunks to return to eat in peace). Tables are tiny and all garbage and food debris go on the floor. The noise is astounding. When mealtime is over, the place is cleared, chairs are piled on tables, and everything is hosed down. The bones and other debris are washed into the river. Everything sits, dripping and drying, until the next onslaught.

Today's lunch is a hunk of very bony fish, some pork and hot green pepper, pork fat and braised cucumber, and egg and tomato soup. This is the biggest and best feast we will have on a boat. Other times, by the time the lines have gone, only pork and large slices of ginger (handfuls of them) remain. But still, a meal costs only about ¥20, $5 for five. Robert says he has eaten worse on ferries.

Two men fish for silver carp. One of them climbs the ladder on the front of the sampan and leans far out, rhythmically raising and lowering a net, untangling a fish and throwing it into the boat. The other man rows, one hand on an oar, the other on the rudder. The loudspeakers play insanely on. The sky clears, the sun burns its way through the murk. The greens intensify. Typical Sichuan farmsteads are scattered here and there among banana trees, sometimes collecting into villages, most often, we think, above the reservoir line. The valley narrows here. Terraces, where terraces can be constructed, are planted with vegetables. The growing season for rice is behind us, but this land can be farmed year round. Water buffalo graze, there is no ploughing now.

2

THE STONE FISH

"Ancient Chongqing's Glory is Fuling's Stone Fish"

(Caroline, on the boat to Fuling, September 16, 1992) We are sprawled on our bunks. *"Xiaoxi,"* small sleep, afternoon nap, explains why there are bunks not benches on these ferries. Kailin translates from a book we have bought on the history of Fuling. In earliest times it was a Ba Kingdom. On the high crag nearby (we discuss the difference between a crag and a cliff), it is said one can see wooden coffins. But it is Baiheliang, White Crane Ridge, that is most famous.

The ridge is an outcrop of hard sandstone, 1,600m long, parallel to and protecting the embankment below the city at low water. At most, two metres of this ridge are exposed during a very low water period, creating what the local people call "Mirror Lake" in front of Fuling. A softer shale stratum below the sandstone is being undermined by the river and the carvings are subject to erosion.

Theorists such as Karl Wittfogel, commenting on how and why some early peoples and not others developed complex societies, have gone so far as to say that water control is both a precondition and an impetus for the development of what we like to call civilization. China is one of their most compelling examples. Yu the Great, the mythical hero said to be the first emperor of the Xia Dynasty, diverted the Min River, as it emerged from the Tibetan plateau, and thus irrigated the fertile Sichuan Plain. Yu was then given the *Treasured Volume of Yellow Silk* on Wushan Mountain by the Goddess Yao Ji, which told him how to dig the Three Gorges to set the surging waters free . . . . (Other more prosaic sources say that what is called the Dujiangyan Irrigation Project was begun in the third century BC.)

But neither Yu nor feudal technology could control the Yangzi. The best they could do was to monitor it. Over the centuries, Fuling has been the most important place for the study of low water patterns for this part of the river. Low water marks were not required for flood control, but for navigation of the shoals. Before descending the Gorges, boatmen would gauge the height of the water from the carvings of stone fish on the face of Baiheliang.

Only in the 20th century have systematic records been made and preserved. How high is a hundred year flood? There is little consensus. One task, set for the team from the Yangzi River Basin Planning Office and the Chongqing Museum that studied these low water marks in 1972, was to determine both the modern and historic low water levels for the Fuling area. The most famous low water mark is an engraving of two carp swimming upstream: the first with a leaf of the lucky ming grass in its mouth; the one behind with a lotus blossom. They are each about a metre long and half a metre through the belly. In 1685 Fuling's governor Xiao Xingong noted that the two fish were eroding and ordered a stone carver to restore them.

The authenticity of these carvings bears on the accuracy of the water levels they record. In AD 971, Xie Changyu wrote in the *Taiping Huanyu Ji* that in 763 the low waters revealed two stone fish which people took to be omens of a good harvest. Are these the fish that Xiao had recarved? The survey team thought so. But under, and thus older than the 1685 "recarved fish," is a barely discernible smaller fish inscribed in official script "*shiyu*," (stone fish). The researchers took the bottom of its belly to be the historic (average) low water level for the Fuling area for measurements prior to 1685. For "Fish in the water at a depth of so many *chi*," they took the fish's back as the starting point. A *chi* is a third of a metre. Many marks are simply: "The water level was here: date." Of 163 inscriptions on the ridge, 103 had useful information for hydrology. Most were from the Song, although Tang inscriptions are known from written records. For 72 of 1200 years, from AD 763, inscriptions indicate particularly low water, with extremely low water every few decades.

These low water marks are considered to be national treasures of China. How are they to be preserved? When the reservoir rises, they will be permanently under water and silted up. An underwater viewing facility (or even small submersibles) have been suggested. A less dramatic but more likely solution is to cut the inscriptions out of the ledge in great slabs and move them to a museum. The Chinese government has said it plans to ask for international technological assistance. The CYJV report rather improbably lists tablets from the Three Kingdoms and the Southern and Northern Dynasties, and says that ¥150,000 is allocated to move them.

The 1972 study team also looked at marks at Chongqing, Yunyang and Baidicheng. Lonji Rock, "Dragon Spine Stone," is near Zhang Fei Temple, south of Yunyang County Town, one of our stops downstream. County records cite a custom whereby young men and women would make a pilgrimage out to the rock, in the third lunar month, to sacrifice a chicken to ensure a good year. Local boatman sing a song that says that the rock grows or shrinks, to help or hinder human passage. An inscription on the rock proclaims: "Ancient Chongqing's righteous glory is Fuling's Stone Fish and Yunyang's Lonji Ridge." The earliest inscription is from AD 1088. All read: "The water level reached here. Date." Fifty-three record lowwater years.

Baidicheng has two low water stone tablets on Small Goose Hill (*Xiaoyanyudui*) at the entrance to Qutang Gorge. The "historic and famous" Large Yanyu Hill was blasted to smithereens after Liberation as a hazard to navigation. One tablet, at 72 m, says: "April 25, 1796, still over eight *chi* submerged." The year 1796 caused some excitement as it is also recorded at Fuling. February and March are the low water months and so this date is unusual. A second tablet was made or remade in 1937, but was too deeply submerged to be studied by the research team.

Chongqing boasts the earliest low water marks. Ling Rock extends out 200m from Chaotianmen, once perhaps dividing the Yangzi and Jialing rivers. The "Xixi Tablet" at Chongqing, from AD 407, is mentioned in the 1254 inscription of Liu Shuzi at Fuling, and on Yunyang's Longji Rock. The Ba County Gazetteer, however, mentions an inscription on Ling Rock from 27-57 AD. In the 12th century, one Chao Gongwu came to view an inscription, which had been mentioned by Zhang Meng in the Tang Dynasty, but he was unable to see it . . . .

The investigative team spoke in 1972 to old boatmen who said they had seen it and pointed out its location, but it was too deeply buried by mud and silt for them to find it. When it appears every "several decades" the Gazetteer says, Chongqingers take rubbings from it. The research team was, however, able to copy some inscriptions from Ling Rock's face. The watermarks along the route would have been known to boatmen and passed from one generation to another. The hydrologists' fascinating task is to correlate all of the clues, from the archaeological and historic records, and from the reminiscences of old timers.

As a sport scuba diver marooned in landlocked Chongqing for a year, I had often thought longingly of the river. Rivers in Canada, where voyageurs' canoes have overturned, yield archaeological treasures. However, the Yangzi is too big and too fast, and every year a tremendous volume of water and gravel scour the bottom. But I did not know then about the Xixi Tablet. Perhaps there would be a project for a team of underwater archaeologists, before the silt gets too deep. They would need a dredge . . .

We will see none of these rocks and ridges as the water is relatively high in September. As we near Fuling, the attendants mop the food and other garbage off the floor and straighten the sheets, towels and pillows for the next occupants of our cabins.

In 1897 the intrepid Victorian traveler Isabella Bird, on approaching Fuling, noted that the river that joins the Yangzi here is variously called the Honton or the Fu or the Wu: it "passes for much of its course through a rich and fertile region, and through a country which produces large quantities of salt," as well as bisecting the vast coal fields which underlie central Sichuan. Fuling was then a great junk port: ". . . with only two portages, goods from the Far West can reach Canton, and as affording, with its connections the Yuan Ho and the Tunting Lake [Dongting Lake], an alternative route to Hankow, by which the risks of the rapids [the Gorges] are avoided" (1899:4).

She writes of "deliriously" landing on green and flowery shores, above the submerged boulders and treacherous summer whirlpool, in her peculiar "twisted stern junk." Beyond, lay what she claimed to be the most picturesque city on the Yangzi, built on ledges of rock, tier upon tier. A fine pagoda crowned a height nearby, and above the town wall, she spied large temples in commanding positions, literary monuments and fine public buildings. However, even she dared not enter the town as anti-foreigner feeling in 1897 was running high.

But prophetically she notes that the river gate is only eight feet high and there is an air of neglect and decay ". . . old Fuling is almost gone." Alas, Fuling today is a dismal place and nothing dispels this first impression. Coal and cement dust befoul everything, and while we see no desperate poverty, people have that small town look of nowhere to go.

The Rich Brown of Iron Oxide

(Robert, On the Yangzi below Chongqing, September 16, 1992) The Canadian Department of Transport or the Coast Guard would not approve of the passenger embarkation provisions on the Yangzi. I suspect, in fact, that they violate Chinese regulations but, as I am to find many times, the China of the 1990s isn't much of a police state when it comes to enforcing things like safety laws.

Because the water level fluctuates so much from one season to the next, the actual piers where the river boats tie up resemble the tidal docks in oceanside ports. They float up and down. There is usually a main platform which, in the case of the ferry terminals in the major towns, is two or three decks high and almost as big as the boats themselves. But the muddy shore in many places slopes gradually toward the centre of the river. That means that the landing platform must be anchored quite a distance out from the bank. Between the main platform and the shore there are often one or more smaller pontoons. These are connected by aluminum walkways about a metre and half wide. These might have rails, or they might not. However, the initial run from the river bank to the pontoons is almost always made on a wooden plank that is seldom more than 30cm wide.

Being in the surge of people pushing to either get on or off a boat is a genuine challenge. The consequences of falling off any of these walkways into the swirling water that is flowing at several metres per second is not pleasant to consider, leaving aside the fact that the river is a sewer. So it is from that vantage point, looking up when I probably should be looking down, that I have my first close look at a Yangzi River steamer. As with most things I see in China, the boats look a good deal better from a distance than they do close up. There are some luxury passenger ships, with names such as *Great Wall* and *Zhaojun* (named for a beautiful daughter of the river whose story I will hear later in the trip). But the vessel we first board in Chongqing is neither freshly painted nor named. *Boat Number 50* has once had a light-green hull and white upper decks, but the rich brown of iron oxide predominates now. Quite aside from its appearance the boat is a complete, total, wonderful zoo of humanity. For the next twelve days Caroline and I are to be the new attraction.

"It is like being an animal in a cage," Zhongling says after our first experience of being looked at curiously by clutches of waiting passengers.

"Does this bother you?" Kailin asks. I think she is a bit embarrassed by the attention given us by her fellow Chinese travellers.

"Not really," I say. "After all, we've come here to look at China so I guess it's fair enough if the Chinese have a good look at us."

The passengers on the boat are cosmopolitan compared to the people we'll meet in the countryside in the next two weeks, where there is even less familiarity with foreigners. But people's interest is genuine and always friendly, and I get used to having inquisitive heads peering over my shoulder as I write in my notebook. If my strange writing intrigues the Chinese then my drawings seem to delight them. When I hold up my notebook for the circles of onlookers to peruse I am always rewarded with nods, approving smiles and comments that I take to be positive.

Downriver from Chongqing there is a relatively narrow section of the Yangzi that is sometimes referred to as the Fourth Gorge. This constriction in the river causes water to rise quickly in Chongqing during the worst years of flooding. Like a painted scroll the misty hills pass by our cabin door. In most places the rocky slope rises sharply from the water's edge. On these hillsides there are garden lots, banana trees and farmsteads with low walls around two or three buildings. Occasionally there are factories and mines. Spindly trees, with all their lower limbs gone, are sparsely scattered over the landscape. Behind the first row of hills there are successively higher ridges, with the last finally swallowed in the clouds. I have been told that between Chongqing and the Three Gorges the landscape is not terribly interesting, but in my opinion, this is not the case.

My imagination and camera lens are instantly captivated by the traffic on the great river. There seem to be least three levels of vessel sharing the waterway. There are the big passenger ships, such as *Boat Number 50*, which I estimate to be about 80m to 100m in length and perhaps 15m in width. It never ceases to amaze me how many of these there are. There are times when two or even three are within sight of each other, passing with flourishes of water and hand-waving, or racing to be the first into a narrow part of the river. Somewhat smaller than these are a class of medium-sized barges, many with open holds and stern-cabins. They range from 15m to 30m in length and, although some are steel-hulled, they are generally built on the same traditional lines as the older wooden junks. In this medium range there are also numerous tugs, ferries and work boats, as well as some tankers. Some of the barges are self-propelled while others are lashed together in rafts of two, four or six, and pushed by tugs. Finally, there is a bewildering assortment of small craft, most of wooden construction.

Judging by the noise and the amount of black smoke enveloping most of the larger vessels I would assume that virtually all are driven by gasoline engines. Most often the fishing boats and other small craft are not mechanically powered, but are rowed. The arrangements vary, and include solo rigs, with one side-mounted oar and a long trailing steering oar operated by one boatman, two side-mounted oars with a rower at each, and triples with one person in the stern and two working far forward. The rowing is invariably done from a standing position.

One can see how closely life on the shore is intertwined with the pulse of the river. Within the city of Chongqing there are numerous boat yards and floating dry docks along the water. Vessels are being built and repaired in facilities that are fairly conventional by Western standards. A couple of days later at Fengdu, I see a marine railway. It consists of a set of tracks several metres apart that run from under the water up a fairly steep rock slope. A frame mounted on railway wheels can be run down into the river and fairly large boats floated into it and secured. The frame and boat can then be winched up out of the water and pulled to the top of the slope, level with the main street of the town. When repairs are complete the boat is lowered back into the river. Similar arrangements are fairly common around the world.

But in many places boats of all sizes are perched up on the sides of the river in most unconventional ways. A couple of full-sized ships, comparable to the one on which we are travelling, are high and dry on a rock outcrop right in the middle of the river. These don't appear to be regular shipyards, but convenient places where the vessels were purposely stranded when this season's flood abated. The shipyard was brought to the ship rather than the other way around. The workers who swarm over the vessels are literally camped out.

The cargoes of the river boats are almost as varied as the sizes and shapes of the boats themselves. Some are loading sand and gravel from exposed bars and estuaries. Others are piled high with bags of salt or chemical fertilizer. Coal is being dumped into the open holds of still others by seemingly endless human chains of porters. I see some specialized tankers lying alongside chemical plants or refineries, taking on cargoes through long pipes extending down to the river's bank from facilities above. In a few places only, there are large cranes that move, in each bucket load, what it might take fifty porters an hour to carry.

Increasing navigation by larger vessels is one of the aspirations of the Three Gorges Dam proponents. The implications of that for employment, community development and local economic conditions is potentially great. What will these changes mean to these communities and their people?

Small Town Hotel

(Caroline, Fuling, September 16, 1992) Fuling, home to 80,000 people at one count, will be drowned. The area is, however, destined for major development. The Sichuan provincial government has promised that resettlement monies will be spent on a "new type of new economic zone" with tax and duty concessions. Along with Wanxian and Fengdu, it is to have a bridge across the Yangzi, improved highway connections with Chongqing, and a modern telephone system. Funds will be available for projects in chemicals, basic pharmaceuticals, light industry, metallurgy and building materials, as well as for tourism and the proposed Three Gorges University.

Eagerly we push and shamble off the boat with all the others, over the gang planks, and up eight flights of stairs. It is just 3:00p.m., but in the narrow passageway that forms the street above, the floodmark from 1981 is lost in shadow. These steps and the lower parts of town have been flooded many times. The large crowd of peasant porters, in blue cotton "Mao" jackets and many with bare feet, soon give up on getting our business. They form a friendly crowd around us, grinning, chatting, smoking, spitting and gawking at the foreigners struggling to take a picture of their water mark.

The slanting sun has by now all but disappeared in the narrow streets above the river. Like most Yangzi small towns, Fuling has no central square; its civic and commercial life is oriented to the river. The ramp-like roads are scarcely wide enough for trucks and buses and are too steep for bicycles. Almost everyone walks. Stone stairways leading straight up offer a shortcut for the physically fit. Thus, our little band turns left and then follows the switchback around to the right. Kailin says we are on Qilin Street, named for the unicorn, a benevolent mythical creature that protects children from evils spirits and brings sons to childless families . . . . It is a good sign.

Li is far ahead now, asking everyone she meets directions to the best hotel in town, and we trudge after her, trying hard to keep up. Finally, we come upon the Zhongshan Hotel, next to the Bank of China. It is quite new, meaning that it is less than eight storeys high and in the "modern international style." The workmanship is crude and the maintenance worse. Wires hang from the blotched acoustic ceiling tiles and the cavernous lobby is dirty. We are offered second best rooms, in the second best wing. Zhongling explains that there are important delegations from Chongqing and Beijing because of the dam, and they have priority. A new minibus with Chongqing plates, full of men in suits, with briefcases and cameras, is parked outside. No problem. We agree to two rooms with bath attached, ¥20 for foreigners, ¥10 for Chinese. Robert, Zhongling and I make ourselves at home on the couches in the lobby with our pile of luggage, waiting for the rooms to be cleaned. Kailin and Li, armed with their letter of introduction, set off to find the local Keeper of Relics to ask him to show us archaeological sites tomorrow.

This is our chance to do the shopping we had no time for in Chongqing. First Robert and Zhongling go, instructed to get some large enamel cups. They come back with two small cups with lids. Fine for tea or brushing your teeth, but too small for lunch! So Robert stays and Zhongling and I go. The little store sells radios and tapes and snack foods and housewares. There are no plastic lunch boxes like Zhongling's (with a compartment for plastic chopsticks) and no old-fashioned metal ones like Kailin's. The best we can do is a plastic refrigerator box. This is a precaution against hepatitis, A or B or C, I am not clear which, and a way of avoiding the styrofoam throwaways. Kailin has said, "Hepatitis A is very popular in Chongqing." The idea is never to trust the dishwashers. In fuel-short China dishes are washed in cold water, straight from the tap. The guidebooks

warn of "hep" sticks, improperly washed bamboo chopsticks, and so the
bowl idea sounds as if it can do no harm. Throwaway chopsticks in restaurants, along with the styrofoam lunch boxes, are a public health innovation
of frightening ecological implications. The traditional "bring your own" is
much preferable.

And so, from now on, whatever the claims of the establishment, we
will pull from our growing bundles of luggage our plastic boxes, tin cups,
and the increasingly grubby throwaway chopsticks saved from the last
place. No one takes the slightest offense. At the end of each meal, Robert
and I will help ourselves to a little boiled water from the restaurant's tea
thermos to rinse out the bowls (having first wiped out the grease with our
facial tissues used for napkins). First the debris, then the greasy water will
be poured into the spittoon or the gutter, depending on the floor the restaurant is on! Robert and I never quite get used to this, but Zhongling, Kailin
and Li are adamant. We do break the rules from time to time, drinking beer
out of the glasses offered, or unable to refuse a friendly cup of tea. But no
one gets hepatitis, so it works!

Chinese people will only drink hot boiled water, a sensible precaution,
but our friends rinse their lunch boxes with cold water from the nearest tap.
They also use tap water, including river water from the tap on the boat, to
brush their teeth (being careful not to swallow but to spit it out). They drink
directly out of a can, without washing off the top. When we point out that
again they are probably rendering their precautions worthless, Kailin
shrugs and then agrees. One can always learn something new!

A stall across from the hotel is selling black cotton strapped slippers,
soled with many glued and stitched layers of cloth. Only old peasant women
wear them today. I buy a pair and put them in the shopping bag. Shopping
in China is like shopping many other places. Do not count on finding something again.

Kailin and Li return successful. A man from the local Bureau of Relics
will come to the hotel tonight to finalize the plans for tomorrow. Our rooms
are finally ready: two painted concrete boxes off a canyon of a hallway. *Xiaojie! Xiaojie!* Holler all you will: the lively teenage keylady, is never at the
desk and, except for tea water, there is no pretense of service. Everything is
fine, however except that the toilet has no seat. I insist. The cracked-off seat
is found against the wall behind the door, and placed on top of the bowl. My
question, "How do you [females] use a [western] toilet without a seat?"
goes unanswered. Toilet seats are of no concern to the staff and little to our
Chinese friends. Triumphant, I use my limited knowledge of toilet mechanics to get it to flush.

It is best, Li says, to eat in the hotel. The dining room, empty except
for us, is upholstered in red velvet, as pretentious and shabby as the lobby.
The men in suits are eating elsewhere. The waitress is eager to please but
little on the hand-written menu is available. Kailin negotiates the ever reli-

able pork with Sichuan vegetable (pickle), chicken livers, braised winter melon with garlic, bamboo shoots with slivers of mushroom, and egg and tomato soup. Robert orders beer and we order peach drinks. Farther up river, only syrupy cola is available and so I switch to beer. Sichuanese do not usually drink tea with meals. This is our first experience with the plastic bowls. We do it.

It is an exuberant meal. After many uncertainties our expedition is well and truly launched. Kailin describes the chilies of Sichuan, the heart and soul of Sichuan cooking, and a source of great local pride. First and foremost is *huajiao*, the red flower pepper corn that burns and numbs at the same time. This is the taste most difficult for foreigners to acquire. Then, there are black and white peppers, *heihujiao*, and the small hot green peppers, *daqingjiao*, everywhere served with pork and Sichuan vegetable. *Haijiao* is ripe just now. But the hottest is the seven-star pepper, a fiery bouquet which ripens in late summer and is hung to dry. It is ground with a pestle and perhaps a little *huajiao* is mixed with it.

Chilies are often added to dishes in oil. Vegetable oil must be heated in the wok until it smokes; only then do you toss in the ground chilies. The smoke, full of chili, travels up the nasal passageways to the lungs. I have seen cooks in their homes coughing their lungs out. Each area has its own special variety and way of mixing chilies and it is not unusual for people to travel with a packet of their favorite blend or to write home for some. Chilies are believed to be an antidote to the damp and cold of Sichuan's winters.

The talk turns to snakes. Kailin reminisces about her school days in Canton at the Foreign Languages Institute in 1978. She and an Australian couple would buy one or two snakes every week for soup. It is important to choose the snake while still alive, hanging from the cut branch of a tree in the store window. The snake merchant decapitates it and peels back the skin, like taking off a kid glove. Cooking snake in the kitchen is taboo; you must do it outside, so the ashes from the ceiling will not fall into the pot and poison you.

After supper, we have another meeting. Li warns us about what we will see tomorrow. In the back country, people live like "wild grass." There are few roads and they are far from schools. Besides farming, peasants sell construction materials: brick, sand, gravel, and cut stones. Life will get better for them, as demand for construction materials grows.

Peasants losing their land because of the dam will be compensated, just as those in the path of the new Chongqing-Chengdu Highway are being compensated. Even if their house is very old, families are given a new low-rise apartment . . . . Li speaks from firsthand knowledge, having done salvage archaeology in the area. When relics are found, archaeologists are called in to take charge. Li is anxious for us to understand that the compensation offered there is more than fair.

There is a controversy over the Red Pagoda which we saw on approaching Fuling. "In ancient times," as Kailin is fond of saying, famous men wrote poems there. There are also inscriptions useful for the study of Yi-Jing. The city government, anticipating the tourist opportunities of the Three Gorges Dam, drew up ambitious plans for a park with ten famous scenic points, including the pagoda. However, the Horticultural Bureau is now slated to control the park and the museum has stopped putting money into the reconstruction of the pagoda until an agreement on splitting admission fees is made. Until very recently everything was decided on high. Now the local people are taking more control and the process is not always harmonious.

Li returns: our contact has come and gone and they have arranged everything. It is "about an hour's drive" to the place where there are important Ba graves. As a van would cost ¥180 (US $45), they have decided that we will save money by rising at 5:30a.m. to walk half a kilometre to the bus station, take a local bus to the town, and then walk to the site. (All this carrying our luggage, I wonder?) We must be back in Fuling to catch the only ship available tomorrow at 11:30a.m. While Robert and I are grateful for the concern for our pocket books, this plan does not sound too good to us! We try to explain to Li and the others that we must make decisions collectively, taking everyone's ideas and needs into account. But, alas, the friend has gone home. All we can do is go to bed immediately (it is 9:00p.m.), meet him as agreed, and ask him to look for a van.

Every TV in the neighbourhood is blaring. Popular, but not the only choice, is a Gong Fa gangster movie, with a Miami-Vice sound track. As Kailin and I finish in the bathroom, Li curls up, her arm over her eyes. She is a pro. I stretch out on the terry sheet. The bed is lumpy but it doesn't sag. Eventually we all drift off to sleep.

3

FULING

Who Cares About the Bellicose Ba?

(Caroline, Fuling County, September 17, 1992) The roosters wake me at
2:00 a.m. After a while, they wail like coyotes. Then it is 6:30. Li, the silent
early riser, has gone to meet her new friend, to ask him to find a van. Wang
Changchun's title is Office Worker, in the Relics Management Bureau of
Fuling County. His colleagues offer to get us boat tickets to our next destination, Zhongzhou. Li and Wang return with a minibus, a *kangfulai.*

We probably took a local bus out of service. The driver has his girlfriend or wife as helper. An athletic and gregarious woman — trou-
bleshooter, negotiator and business agent — she leaps about, leaning out of
her window to talk, reaching back and closing the door. It is a typical half-
sized bus, built on a truck chassis, with a large flat area behind the driver
for access to the motor. It is hot, as we find out, having set our film bag on
it! There are few springs and the scantily padded seats are fixed, so the ride
will be rough and noisy.

Fuling is no more attractive in the morning than it is at dusk, as people stoke their coal braziers to boil their *xifan* (rice porridge) and noodles,
and the factories start up. People are washing, spitting, doing *taichi*, running, squatting and eating, all by the roadside in the smog and gloom. We
set out in the sulfurous mist, south through the Wu River Valley. There will
not be enough light to take pictures for a long while yet. The bus skirts a
construction project and pulls up at a toll gate. This road, a tremendous
achievement, has just opened. We are told it has been under construction
for thirty years. Otherwise the only way into this back country is via the Wu
River that rises in Guizhou Province.

The road is just one lane and a bit wide. Its crushed limestone surface
is rough enough to shake out our fillings. Picture-taking and note-taking
are impossible. When we stop to look down at the remains of an ancient
plank road cut into the cliffside far below, we realize that the road bed sits
in a half circle hacked out of the hillside, very much like the old trackers'

paths. There are few barriers between us and eternity. The view from the river side of the bus is exciting. Forget those stories of busloads of Taiwanese tumbling over cliffs in western Sichuan! The river bank below is a hell of smoke and dust from cement factories and coal mines, the ugly reality of industrialization in this otherwise green and lovely area. Barges shunt back and forth. Who would have guessed so much activity on a tributary?

A lone man makes gravel in a quarry by pouring rocks, basketful by basketful, into a small gas-powered crusher. This quarry boasts horse drawn carts. Horses are seldom used in Sichuan: people walk, carrying their burdens on their backs. Men pry boulders from bright yellow veins, shape them with hammer and chisel into blocks, and pile them on the roadside for pickup. Small brickworks are everywhere, belching smoke from their coal-fired beehive-shaped kilns. An old man stands up to his knees in the muck, forming the bricks in wooden moulds and piling them in rows to sun dry before firing. These are the township enterprises Li described. Small, brutal, inefficient, with minimal equipment, without any seeming regard for the environment, but thriving.

The sprawling one-storey rammed-earth or brick and concrete farm houses are traditional to old Sichuan, but there are more palm, banana and bamboo trees than one sees near Chongqing. These bananas do not bear fruit: their leaves are cut to wrap festival treats of sticky rice. Tiger lilies bloom in the wild grass and weeds beside the road. Mothers appear, suddenly through the morning mist, bringing their brightly-bundled children to the roadside, presumably to wait for the school bus. We are mindful of Li's warnings about the "backwardness" of these people. Although the farmland is high and poor and the hillsides are covered often only with scrub and bush, our impression is one of industry and activity.

Soon we are climbing up and over a high pass. The cliffs on both sides rise abruptly from the river, twisting far below. Neither trees nor dwellings cling to these rock faces. We pass through a small tunnel and then descend. The valley opens wide and switchbacks bring us down into brilliant green bottom lands where small tributaries, low now, enter the Wu River. Each stream is spanned by a finely-engineered stone bridge. There are paddy fields at the lowest levels and terraces stretch as far up the mountains as crops can grow. We catch a glimpse of a woman in a farmyard weaving a bamboo-strip fence to keep her chickens in.

In much of the Yangzi Valley, the rising water of the reservoir will cover rock faces and deforested hillsides. But these tributary rivers will almost stop flowing as the water backs up into their valleys. It is here that productive farmland will be lost, including the valuable rice paddies that have been won with so much labour over so many centuries. Other valleys, now locked in the mountainous interior, may of course be opened up. The maps we have show no towns or roads or even major rivers through large areas north and south of the Yangzi. We can get no idea of the number of people who might live in these areas, although if we extrapolate from the Wu Valley, it must not be large. Peasants, now carrying out their produce on their backs over stone pathways and steps to the rivers, will have easier access to the cities as the infrastructure improves. The reforms of the past ten years have brought little benefit to those locked in the far countryside.

Finally, we reach another toll gate. The road ends in a straggling village, the first we have seen in an hour-and-a-half of hard driving. The valley is dominated here by an immense new fertilizer plant. It must also mine potash, as an enormous chute snakes up and up, far back into the mountains. Xiaotianxi Site is famous for Ba Kingdom graves, dating from the end of the Warring States Period (400-200 BC). In 1971, a child found bronze "kettles" in a grave near a pigsty. The peasants did not know what they were but, luckily, reported them to the municipal authorities. Excavations were carried out in 1972 by teams from the Fuling County Cultural Centre and the Chongqing and Sichuan museums and reported in the archaeological journal *Wenwu (Relics)* in 1974. They found three Ba Graves in a small gully running down to the river. Further finds were made in 1981, 1984, and 1990, indicating a significant concentration of high status burials here during the Warring States.

We make a three-point turn, at an incline of 45°, and then bounce along a track, past abandoned gray brick buildings with fading medallions under their eaves of Chairman Mao wearing his Red Army hat. Were these buildings not needed when the communes gave way to the household responsibility system? There are few reminders now in the countryside of Mao or of the Cultural Revolution that was intended to bring these remote villages into the mainstream.

Breakfast in the village has been promised. We stop at a school and a line of empty shops and the driver's helper negotiates with the local people who have come to meet us. Rice porridge and steamed bread, *mantou*, will be ready on our return. That is all they have. Anything will do! The sun is out, it is 9:00 a.m., and we are hungry. Already we are running behind schedule. Only ten minutes to eat! But first, a bathroom break. The "ladies" is next to the pig pen and above a cesspool. A half-grown pig snuffles and looks calmly at me, its nose only a few inches away.

The Ba, and the more westerly Shu, are the two early tribal peoples who dominated what is now Sichuan. Most experts agree that their cultures were identical, the distinction between them being mainly geographical. They have now been identified in archaeological remains of the Late Neolithic and their culture flowered at the time of the Xia and Shang dynasties. They have a reputation for fierceness and their remarkable bronzes are so far unique in Chinese history.

By 700 BC they had formed chiefdoms or early state societies: historical sources speak of a "Ba Kingdom." But, in 316 BC, the Ba and the Shu were defeated by the Kingdom of Qin and their chieftains became vassals,

first of the Qin and then the Han. They were garrisoned in Sichuan to keep
the more barbarian barbarians in check. Remnants of the proud Ba re-
treated to remote areas of southern and eastern Sichuan. Their descen-
dents in the Three Gorges, according to the tourist books, have distinctive,
(and, in Ba tradition, contrary) customs: the bride weeps in her wedding
procession and mourners at funerals beat gongs and drums and sing.

Interest in the Ba and Shu ancestors is high in scholarly circles, and
as a source of pride and of inspiration for local artists. In the 1980s, Sichuan
Television made a series on Ba culture, claiming them to be the true ancestors of the people of Chongqing (the center of the Ba, whereas the capital,
Chengdu, claims the Shu).

Ba-Shu bronzes have been found everywhere in Sichuan (but not in
the mountainous northwest), and in neighbouring provinces. Archaeologists have used them to distinguish the usual three periods of development:
early, middle and late. Generally there is a progression from striking originality in earlier finds to virtual assimilation by Han times. The first period
corresponds roughly to the Shang Dynasty and the most spectacular site
from this period, or any other, is Sanxingdui, Three Star Mound, at Guanghan, north of Chengdu. It is thought to have been the site of the ancient
capital of Shu. Sixteen cultural layers, distinguished by their pottery types,
indicate continuous occupation from Late Neolithic (prebronze) to Qin
times. Its discovery, in 1980 by local brickmakers, is one of the most sensa-
tional finds in recent times, ranking with Schliemann's finding the "Mask
of Agamemnon" and Woolley's excavation of the Treasury of Ur. The dream
of other such sites drives the search for more of the Ba and the Shu.

Two sacrificial pits yielded over a thousand bronze, ivory and jade
pieces. They include the largest bronze statue yet found in China; the
largest bronze mask; 54 highly-stylized human heads, some with flaring
ears, large noses and jutting dowl-shaped eyes; and mythic tigers, dragons,
snakes and birds. A bronze tree, almost four metres high, was laden with
fruit, exotic animals, bells, and plaques. Archaeologists also found a solid
gold mask, a staff covered with gold foil, and many small gold foil objects,
the oldest gold relics yet found in China.

Who, then, were these Ba-Shu? Their style is utterly unique in Chinese art, apparently appearing suddenly. To an outsider, it has more in
common with the art of the islands of Southeast Asia than with China.
Some Chinese writers, however, see their animism and worship of holy
trees as possibly coming from West Asia, although tree worship was common in Shang times.

The second Ba-Shu period spans the middle years of the Warring
States Period, around 300 BC. At Majia, in Xindu County near Chengdu,
the 180 bronze vessels found in a wooden coffin are more like those of the
Kingdom of Chu and of the Central Plains.

The third period spans the Late Warring States and the early Han

Dynasty. It was a time of defeat and submission to the culture of the conquerors. Its most spectacular relics were found here at Xiaotianxi: a set of 14 bronze chimes, *bianzhong*, inlaid with gold filaments. But two bronze drums, a *chunyu* and a *zheng*, are perhaps the most important.

The warlike Ba-Shu made innovations in bronze technology and in bronze weaponry. Their *ge* and double-sheathed swords are unique, as are their "pouch-shaped" spears. Some bronze *mou*, *fu* and *zeng* vessels are believed to have originated with the Ba-Shu and then spread to Qin and Chu. These vessels are still integral to Chinese cooking and ritual.

Few Ba sites have been systematically excavated and so these largely undisturbed graves are important. Li says, with professional caution, that while we cannot even say for certain these graves are Ba, the *Hua Yang Guo Zhi, The History of Hua Yang State from the East Jin Dynasty*, records that the Ba came here in the Shang, making Fuling an administrative centre in the Warring States. The excavation report says that although direct proof the occupant of the richest grave was a Ba king is lacking, comparisons with graves found in Dongsunba, a Chongqing suburb, indicate he was an important person.

The graves were generally four by two metres. Grave 1 was the largest, at six by four metres. The walls were tamped earth. Because the soil is damp and acidic, the wooden coffins and corpses have long ago decomposed. The coffins were perhaps two metres long and a metre wide: archaeologists found traces in the soil of red lacquer that must have been applied to both interior and exterior surfaces. Five bronze rings with tiger-head attachments were found in Grave 5 in just the right spots to indicate that they were carrying rings. It also contained a *ge* lance dated from its calligraphy to the 26th year of King Zhao of Qin (281 BC).

However, archaeologists were most excited by the two drums, rarely found together, in Grave 2. The *chunyu* had a tiger-head knob set with black beads for eyes; the Ba apparently adapted it from Qi and added their totem. The *zheng* was decorated with the Ba character "⊲⊳" along with two seal-script *"wang"* characters, meaning "king." Literary sources mention that they were played in battle to command troops to advance or retreat and they indicate that the chief buried here had troops under his command.

The best-known legend of Ba origins tells the story of Lin Jun and five clans who lived on Wuluo Zhongli Mountain, near Yidu, south of Yichang. In the mountain there were two caves: a red cave where a son, Wuxiang, was born to the Ba, and a black cave where sons were born to each of the other clans. The tribe had no chief and so, when the young men were grown, it was decided that the one who found a cave by lancing the rock with a sword would be chief. Only Wuxiang succeeded. But the clans were not yet satisfied, and so the young men were set afloat in earthenware boats. All except Wuxiang sank, and so he was named Lin (Lord or Chief) Jun. After his death, it is said that he turned into a white tiger. The *Hou-*

Han Shu records that ancient Ba sacrificed human beings to white tigers they believed to be reincarnations of their founder, Lin Jun.

In 1972 an earthenware model of a boat, 7.2cm long, was unearthed near Yidu, suggesting to some that it was a ritual utensil used to choose Ba chiefs. The relic was linked to the local Baimiao Culture, which is thought to differ from its contemporary Xia-Shang cultures in western Hubei. Later, legend says, the tribe of Lin Jun moved west to Sichuan. There is also a tradition that, when a white tiger terrorized Yunyang County, a small band of men killed it with a white bamboo crossbow. The same King Zhao of Qin, whose name was found on the sword at Xiaotianxi, rewarded them.

Western scholars divide the language of SE Asia into three major families: Austroasiatic (Vietnamese, Mon, Miao-Yao, Khmer); Sino-Tibetan (Chinese, Tibetan, Burmese) and Austronesian (Cham, Malay, Indonesian languages). The classification of the Tai language, and the homeland of the Austronesian speakers, have been two of the hottest topics of linguistic debate in the past ten years. The level of disagreement indicates the rudimentary level of our knowledge of these languages.

The first Ba characters (a tiger, a hand and a flower calyx) were found before Liberation on Zhanguo-era (400 BC) bronze weapons. They were not part of the ancient seal script and they were not merely ornamentation. Then, in 1954, bronzes inscribed with similar characters were unearthed from boat burials at sites attributed to the Ba at Dongsunba (Xu 1974(5):2025). (These boat-coffins are displayed on the second floor of the Chongqing Museum.) The Ba language and script were gradually replaced by Chinese after their defeat. Ba characters, except for a few pictograms, have not been deciphered, but over two hundred have been recorded from relics and seals found at Sichuan and Hubei sites. Finding more characters, and deciphering this *Ba-Shu Tu Yu*, Ba-Shu symbol language, would answer many questions.

According to linguistics expert E. G. Pulleyblank, "The Ba still existed as a non-Chinese tribal people in southeastern Sichuan in Han times. The *Hou-Han Shu*, the *Book of the East Han Dynasty*, classified them as *Man*, which ought to mean that they were Miao-Yao speakers, rather than Tibeto-Burmans (1983:422)." The Miao-Yao, Tai and Vietnamese languages are similar to Chinese today, in their monosyllables and tonal systems, and they have freely borrowed Chinese vocabulary. However, the related Mon and Cambodian languages, spoken by peoples who have not been in constant contact with the Chinese, lack the tones that have developed in Chinese only over the past 2,000 years, and thus may contain vestiges of the now-dead languages. Recently dictionaries have begun to appear for the languages of central and south China, and a reconstruction of their proto-language has begun that may shed some light on the language of the Ba.

Pulleyblank also cites K. C. Chang's contention that, when the Han Chinese penetrated Sichuan in the Warring States Period, the Ba showed close cultural affinities with the Chu, and the Shu with the Dongson and Tien peoples of Yunnan. The fact that the Chu State "arose during the first millennium BC in the middle Yangzi region, in the midst of the Man tribes," strengthens the argument that the Ba and the Chu were Miao, or austroasiatic speakers, and were later sinicized. The Chu word for tiger, "*Yu-t'u*" in the ancient texts, can be shown to be similar to Mon, Khmer and even Burmese words for tiger. Pulleyblank notes that the Miao tell stories of men transformed into tigers and ethnographic studies show that present-day Miao believe that some human souls after death become blood-sucking vampire tigers. Tigers and tiger-myths, however, are widespread and no guarantee of ethnic affiliations.

SE Asia still has hundreds of tribal groupings who have kept remnants of their cultures despite the attraction of Chinese civilization and the power of China's armies. The Ba-Shu and the peoples of Yunnan and Guizhou are also drawn together by the importance of bronze drums in their cultures. Pulleyblank cautions us, however, that while it is tempting, it is premature to classify these groups together and identify them with surviving peoples (1983:426-35).

Is there an "Austronesian homeland" in South China? Is Tai an Austronesian language? How should we classify the Ba language? Are their strange bronzes linked to the art of East Asia? This is Thor Hyderdal stuff, but the possibilities are endless for those who enjoy historical speculation. Archaeology, ethnography and linguistics will all eventually cast more light on the culture and origins of the ancient Ba.

We jump down from the bus, following Wang Changchun as quickly as we can, single file along the earthen dikes above the sodden paddy fields, and up and down mud-slick stone steps between cabbage patches and sweet potato beds. We make way for two men carrying a large pig in a sling, along the dikes.

Wang stands at the mouth of a ravine leading down to the Wu River, telling (in rapid-fire Chinese) the story of the graves found here in 1971. Finally Robert gets his thoughts organized enough to hand Zhongling the dictating machine to tape what he is saying. Kailin has given up trying to translate and is scribbling notes and trying to keep me from falling, cameras and all, into the muck. As always, we must hurry.

The truth is I cannot make much of the site: Li promises to give us an article on the 1972 excavations. Unfortunately it is of the laundry list variety, listing the extraordinary grave goods, but lacking interpretation. She apologizes that it is typical of the training of archaeologists during the Cultural Revolution. For all the talk of working people, there was little interest or training except in looking for beautiful objects. Pot hunting. They do better now.

Back we go, single file, along the raised paths to meet the farmer, Chen Jiajie, who made one of the later finds. He was rewarded with a payment of ¥200. Robert's tape recording includes Wang's surprising assertion (given the acidic soil) that human skeletons were found, heads oriented to the south. As well, there is evidence, Wang says, that people were buried alive, outside the coffins. The Ba-Shu kingdoms are thought to have been slave societies. Elaborate grave goods indicate to archaeologists that social stratification (class division) is well established.

Desperate to be doing anything of value, I videotape Chen's story and surreptitiously tape some of the people gathering in a silent crowd around us. The video is to give some idea of the logistics of archaeological excavations in this part of China. This village is two hours hard driving from a town offering accommodation except for peasant's houses. The site is in the middle of the village and excavation would require the villager's cooperation.

We are off again, back to the village centre, and then along the riverbank. Here there are undisturbed archaeological strata, with shards of the Warring States and other periods sticking out of them. To see them we have to stand almost up to our ankles in muck, in a cabbage field that inclines at a 45° angle. Down, down go Wang and Li into the small brickworks, an area pockmarked with large pits where clay has been mined. She has her archaeologist's trowel and is helping herself to a shard here and there.

I am beginning to think that brickmakers should be trained in archaeology. They make the finds. A second article, found later in Toronto (*Kaogu* 1985(1):14-17), details their discovery of four more graves (M4-M7) in 1980. Again the coffins and their occupants had decomposed. The 46 offerings included an immense and fine bronze urn, pots and pans, jade, pottery seals and glass beads. These graves were less richly endowed than the earlier ones and did not contain their military trappings.

Now that we know to look, the fields contain literally hundreds of thousands of pottery fragments. But in the bank, which is collapsing bit by bit into the river, they lie in sequence as they were discarded, along with animal, plant and other remains. Experts from Beijing, we are told, have said that they span the Shang to the Han dynasties. Wang would like to dig here, and he thinks that the government will authorize further salvage work. The CYJV report says the altitude is 180m. This depends on where you are standing! The villagers have been told that they will move, and they are getting ready. Sadly, there is no time to talk to them through our interpreters. They grow rice and vegetables: soya beans, peanuts, sweet potatoes, squash, seven-star peppers, red peppers, cabbage and egg plant. Kailin draws on her experience to explain that the sweet potatoes, at this stage, must be "turned over": the extra tubers just now sprouting must be pulled out so the main tubers will grow into large potatoes.

Kailin was a Red Guard during the Cultural Revolution. Then, as an educated young person "answering Chairman Mao's call," she went to live in the countryside in 1969. The first year, she worked in the fields beside the peasants. Then she was promoted to brigade accountant and started to rather enjoy her stay. However she shares none of our romantic views about these sons of the earth and their lot in life.

We do not know much about the daily life of the Ba in this period, how much life here has changed over the millennia. Historical accounts speak of them as hunters, and of the dugout canoes in which they sometimes buried their dead, indicating that they were well adapted to fishing and a riverine life. They practiced agriculture, probably on terraces and bottom lands. Their metallurgy required a high degree of skill and social organization.

The Sichuan Bureau of Relics erected a stone plaque in 1981 declaring the village an historic site, but I gather it "buys" good finds from the locals to stop the illegal trade in antiquities. The treasures of Xiaotianxi are proudly displayed at the Sichuan Provincial Museum in Chengdu. Kailin and Li mention a "showroom" in the Fengdu relics office but we are told it is not ready to visit. Clearly, these people want a local museum.

We have to go or we will miss the boat, lose a precious day, and the hard work getting the boat tickets will be in vain. We retrace our steps through the village's single street. The women put down their knitting as we go by, a girl in a dim and dusty "office" swirls the tea in her cup and yawns. Perhaps she is the accountant here. The Red Guards like Kailin have long returned to the city. I photograph a blacksmith working the bellows at his forge, a water buffalo shuffling in his stall, and a large millstone. Kailin says the small headless statue draped in red cloth is not a Buddha but a local god. Incense sticks have recently been burned in front of it. This village, the graveyard of Ba chieftains, will be drowned.

Breakfast is pronounced too dirty to eat. We will have to wait until we get back to Fuling, a two-hour drive. (After this experience, I try always to have some snack food in our bags and bundles.) Back into the bus and onto the toll road we go, stopping for a few minutes to photograph coal-fired brick kilns, a local industry that will continue in the traditional way.

We are waved to a stop by the local police, now occupying a checkpoint. Quietly we put our cameras down and away, and wait. The bus driver's helper explains, we assume, that these foreigners have been to look at the archaeological site, and goes inside. We are on our way out, we are with the local relics bureau, what can they do? (Later we find out!) We do not know if this is a closed area; we have no permits. Many remote counties of Sichuan have strategic factories, built during the Cold War years in case of Russian invasion, but now they are converting to peacetime products. The woman returns; we continue, and nothing is said.

I imagine the Wu River as a salt route to mysterious Guizhou Province. At one point the river is almost spanned by long semicircular shelves of rock, but navigation still is possible at this time of year, up to our village at least. How else, until recently, could they get the fertilizer out? Raising the water level will extend navigation further upstream. Later I learn that the area is rich in aluminum ore and a huge smelter is planned.

Back we go, down the valleys, over their stone bridges, over the high point, and into Fuling town. The driver leans on his horn, scattering pedestrians and peddlers: we are going too fast for the video camera to focus. We pass every sort of small workshop and provisioner: weavers of bamboo baskets, a shop full of large pottery water jars, a tin-smith, a forge, a ship chandler, restaurants, ramshackle temporary bamboo and wattle houses, and an old river gate, the entrance rising like a tunnel. Will the hustle and bustle of the old port, its merchants relocated, survive?

Below: Two of Fourteen Bronze Chimes and an Immense Bronze Urn found in the Graveyard of the Ba Chieftains at Xiaotianxi, Fuling County.

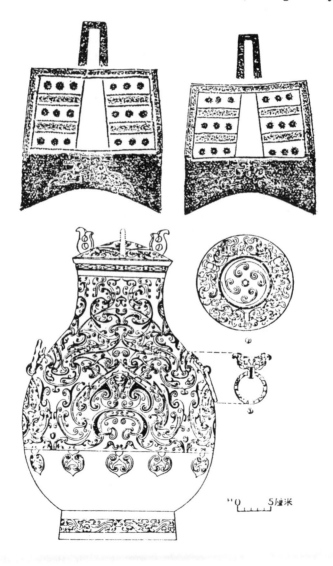

Sanxingdui: Three Star Mound

(Kailin, north of Chengdu, July 23, 1993) My husband brought an article, "Bronze Statues Surprise the World" in the *Chengdu Evening News*, to my attention. Let's go I said. And so we set out, with our son, to investigate the significance of Three Star Mound (Sanxingdui) for Ba-Shu Culture. The tuk-tuk driver took us to Three Star Mound, but the excavations have been finished for now and there was nothing to see but the prosperous farms and the rocky banks of the Moon River. In fact, the mound is the remains of the rammed-earth wall that enclosed three sides of what archaeologists think was the capital of the Shu Kingdom. The city, which has large buildings and palaces, is thought to have covered an area of 12 square kilometres. The Moon River was the "fourth wall."

Next we visited the work station. A group of archaeologists from South Korea had just arrived. Lei Yu, work station director, received us. He explained that in 1988 peasants digging clay for bricks found two pits, known now as Sacrificial Pits numbers one and two. These pits, four by three metres, were first thought to be graves but no skeletons were found. What they did find was one of the largest and certainly one of the most unique caches of bronzes ever found in China.

The bronzes were accompanied by large jade ritual swords, some gold foil ritual items, and an enormous quantity of elephant tusks. They had been placed in the pits all at one time, in layers, in what must have been some sort of ceremony. (They estimate about a hundred-year time difference between the two pits.) Some of the artifacts had been intentionally broken. Were the Shu making a sacrifice at a time of national peril, or were the ritual items broken and buried in an act of defiance against a god who had failed them? The finds are thought to be contemporary with the Shang Dynasty, and refute the view that developments in SW China lagged those of the Central Plains.

Many more practical questions are being asked. The gold content of the foil is 98%. Sichuan is not rich in gold, and thus it must have been imported. Likewise the tusks. Elephants are long extinct in Sichuan. Were the gold and the ivory tribute from tribes to the south? Some of the bronze pieces were welded, an advanced technique for the metallurgists of the time, and some of the pottery vessels with "sharp bottoms" look like modern-day Sichuan pickle pots!

The finds caused a sensation. More research and excavation are needed, as are the funds and modern technology to protect the bronzes from a pernicious "rust" of copper oxide. Meanwhile, the Director is hoping that the relics can go on tour, so the world will see the miracle of Ba-Shu art.

"Dejeuner Sur Deck!"

(Caroline, Leaving Fuling, September 17,1992) Too soon, we are stopped by traffic. Kailin leans out of the window and negotiates a good price for five huge turnip-like things that she says we must try. "We'll miss the boat," I grumble, "and we already have too much to carry. Throw those turnips away!" but she persists. Finally we are out of the jam. The driver drops us off at dockside and we gratefully pay and thank him. Wang's friends rush to meet us; they have tickets, but only fifth class. No bunks, no seats. Perhaps we can upgrade them on the boat. Breakfast is now brunch. We frantically buy up all of the bread and buns from the nearby stalls and pass them around. Our new Fuling friends are hungry too. We thank them and say good-bye.

Thank goodness the boat is late. Kailin and Li have time to check out the restaurant, built with tarps and bamboo poles on the edge of the river-bank beside the stone stairs. The counter is resplendent with platters of fresh green vegetables, washed and chopped, ready for cooking. The cook is questioned and wins Kailin's approval. Toot! The boat is in sight. Kailin supervises the quick cooking of a green spinach-like vegetable (*tentencai*, cilantro) with chili, pork and green pepper, wood ear (dried fungus) and pork, more green vegetable, more pork. Soon everyone in the restaurant-owner's family is busily making a feast. Our bundles are piled on a small table and we have made ourselves at home. In the corner, the TV blares what could be the same Miami Vice-Hong Kong crime show that kept us awake last night.

The ferry is unloading. We pack the remaining food into our plastic containers and join the other boarding passengers on the steps. One man is carrying two huge bundles of empty kerosene cans slung from his bamboo pole. Robert takes pictures of the garbage floating by on the river and of two pigs rooting for dinner in what appears to be the town dump. Garbage is just tipped over the edge of the road, onto the riverbank close to the stairs.

Zhongling suggests a boarding strategy and we soon have it down to a fine art. People push and bunch up to get on (no patient queues in China). We organize ourselves, foreigners first, followed by Kailin and Li. Zhongling, with the tickets, brings up the rear. The idea is to push past the ticket-taker and get on. If the ticket-taker says foreigners are not allowed to travel fifth class, Zhongling will put up a terrific argument while we scurry out of sight and the other passengers bunch up behind him.

We all get on, no hassle. There is no one at the ticket wicket in the passageway, and so we put down our bundles and wait. Meanwhile, others with fifth class tickets are staking their claims to patches of hallway or stair. Finally the agent comes. Sorry! Sold out. I suggest to Zhongling that he try to make a private deal to pay somebody to relinquish a bunk so we can at least sit down. He disappears for a long time and we wait.

Kailin and I agree, as a fallback strategy, that we will sit on the "tail" of the boat, the stern deck. She suggests we rent some bamboo mats before they are all gone, but we can do nothing until Zhongling comes back with our tickets. Finally he returns, no luck. We settle for a mat apiece. The "tail" is covered with garbage left by the previous occupants and thrown down from the deck above. (Falling garbage is a problem during our brief tenancy, but we are more worried about flying cigarette butts.) Kailin and an army corporal, also without a bunk, find a twig broom and sweep the sunflower seed shells and other things into the river as best they can — there is a ledge and everything has to go up and over. Things are quite well organized given the circumstances, but how would a foreigner know about mats and brooms?

The corporal has staked out the area near the capstan against the wall (out of the wind and falling debris) and so we are left with the edge. We line up our mats along the rail, stack our luggage in front of the capstan, and crouch or sit as best we can. The mats are broken but clean enough, especially as compared to the deck.

The first order of business is to finish lunch. A large crowd gathers to stare at the crazy foreigners and their friends, picnicking on the deck. Kailin, who first warns us to keep a hand on our cameras, finally has a chance to take the nail file on Zhongling's nail clippers (Robert's Swiss Army Knife has disappeared and we never do buy another knife) and peel the "turnip." The gray skin comes off easily in large swaths, revealing a crisp white vegetable that looks rather like a huge garlic. No one knows what it is called in English but Kailin translates its Chinese name as *digua* "earth melon." It is crisp, mild and faintly sweet, and it quenches our thirst. We all have some for dessert.

We are well fed, surprisingly comfortable and very proud of ourselves. The trip up the Wu River is pronounced a complete success. The boat ride to our next stop will only, after all, be a few hours. I root out the Nescafe from the shopping bag and get some boiling water from the spigot in the passageway, in my new enamel mug. Everything feels just fine: if only the rain holds off. But Kailin is not so sure and takes off, to find the captain, as we later find out.

We must wait in line to dock at Fengdu, the City of the Dead, a very popular stopping point for Chinese tourists. Fengdu's special festival is in early March, when people come from Chongqing to visit and to burn incense for their ancestors. Everyone crowds onto our piece of "tail" to see the famous sites and we are forced to pick up our mats and defend our pile of luggage. Forty-eight temples are known to have been built on Pingdu Mountain since the 7th century. Li has warned us away from the "Palace of Hell" and the "Boundary between the Living and the Dead." We only have a few days and everything here is well known, a "fake," a tourist trap, a raucous theme park of plump gaily-painted and grinning devils, subjecting their guilty charges to their just rewards. It is Buddhism corrupted by local gods, and even a whiff of sulfur from the Christian Hell.

A sign proclaims the new building not far above the water level to be the Fengdu International Hotel under the auspices of China Tourism. How could it possibly escape the rising waters? The Three Gorges Dam is the first large project to be built in the regime of the "socialist market economy." Some of these buildings are undoubtedly being built with expropriation profits in mind. But the famous temples that line the top of Pingdu Mountain will be safe. The guidebooks put a good face on it by saying that when the water rises, the tourists will just disembark. No more steps!

Kailin returns victorious. She had originally tried her not inconsiderable charms on the second-in-charge, but to no avail. Now she has convinced the captain that this poor (old!) foreign woman should not have to bear the wind, which is rising. He has said we might sit in the second-class lounge. Robert and I protest that we cannot leave our friends. Kailin insists: she has gone to considerable trouble and there is nothing to be gained by our sacrifice. We, who have grown up in relative ease, are not comfortable with the Chinese way of grasping whatever advantage is offered. Our good fortune will engender no envy in our Chinese friends, but much scorn if we turn it down!

The captain proudly leads the way. (We get a tantalizing glimpse of the white sheets of the two-bed second-class cabins as we go by.) The lounge is a bit smoky despite the wind that comes in around the ill-fitting windows, but there are wide red vinyl couches and very few passengers. Of course, our friends are allowed to stay and we set about making ourselves comfortable. At 3:15 we are still waiting. Fengdu's dock area is a hive of activity. A gang of men load white polypropylene sacks of salt onto their shoulders two bags at a time, across a high gangplank and up onto trailers drawn by "walking" tractors. A large wooden-hulled boat, its hull made of logs left round, docks, drops someone off and goes. Old men rinse fishing nets in the water at the foot of the ramp.

We help ourselves to the captain's complimentary tea and eat some of the coated peanuts Zhongling has bought "over the side." Li, the action film addict, has long ago disappeared into the TV room. Just past Fengdu town, the high green bank is dotted with many small family tombs, some decorated with wreaths of paper flowers. A procession winds its way up the hill. These are the family tombs, the tombs of the ancestors of living people, that must be moved or abandoned.

The Yangzi is a busy highway. Disembarking passengers from the local ferry pass porters
unloading cargo from a barge. In summer when the river crests, the mud flats behind
would be covered with water.

Fengdu's shipyard is typical of riverside industries with their ramps and walls. The town
will be largely flooded but the hill in the background, with its popular "City of the Dead"
attraction will remain.

New and old co-exist in urban China. Traditional timber buildings with their wide eaves and tile roofs beside more modern brick structures; exposed water pipes, electrical wiring and open drains.

Locally mined coal fuels an ancient style brick kiln in the Wu Valley. Few trees are left on the mountain side. The building boom promised by the Three Gorges project could benefit such rural enterprise.

Caroline and friends examine exposed neolithic archæological strata in farmer's fields. Even in remote areas there is smog from coal burning industries. Clay for bricks has been dug from the pits below. Near this spot some remarkable Ba culture tombs were found. The whole area will be flooded.

The Many Storied Pagoda towers above the riverside tourist town of Shibaozhai. The town but not the pagoda will be flooded.

The massive nature of traditional Chinese timber construction is evident in the interior of the pagoda at Shibaozhai.

Fu Kailin wanted her picture taken in front of the statue of the legendary hero Ch'ing Liang Yu. When Ch'ing's husband was killed in a Ming Dynasty battle near Shibaozhai she led the army to victory.

From the great stairway up the side of Shibaozhai the broad expanse of the Yangzi Valley can be seen. The wide, fast flowing river will become almost a lake when the Three Gorges Dam is completed.

The traditional wooden junks of the Yangzi are home as well as work place to many people. What will become of these smaller boats, when the new dam opens navigation to larger ships, is not certain.

Mr. Zhen, the archæologist for Zhongxian County, explains to Caroline and Kailin the nature of the neolithic artifacts being found in this site. He has drawn a picture on the ground of the kind of pot from which the sherds came.

This Yangzi tributary above the city of Wanxian is being harnessed by a small scale hydro-electric dam. Local materials and indigenous construction techniques are being used.

Our local guide leads the party to an archæological site in the Gan Jin River Valley. Even in this river bottom land people are growing crops. The stone and gravel which is a valuable local resource will be covered by the waters of the reservoir.

The kneeling woman is burning incense and conducting an I Ching ritual before the fierce faced statue of Zhang Fei.

The graceful upturned eaves of the Zhang Fei Temple stand opposite the town of Yunyang. Both town and temple will have to be moved.

Tourists and local hawkers mount the stairs to the Zhang Fei Temple. In 1870 a "Thousand Year" flood carried away almost the entire temple complex. Again in 1981 the river covered almost everything in this picture. The reservoir will drown the site forever.

The "Poet's Gate" at Fengjie is almost the only remaining fragment of the walls that once encircled all the towns and cities along the Three Gorges. It is probably constructed of tamped earth faced with cut stone.

The Kuimen Gate is the entrance to Qutang, the most westerly of the Three Gorges. Seen from the hilltop of Bai Di Cheng it is one of the most spectacular views in the world.

On the Daning River boatmen still "track" or manually haul their craft up stream.

Small boats along the Yangzi are often propelled by oars alone. Many of these boats are still employed mining sand and gravel from the river bottom.

Some of the buildings in the town of Badong are quite new and have been constructed in defiance of regulations. When they are flooded their owners, mostly local agencies and state companies, hope to be compensated.

Xiling, the last of the Three Gorges... under a cloud or being illuminated by rays of hope?

Fengdu: Ghost City

(Kailin, On the Yangzi near Fengdu, September 17, 1992) Fengdu is dominated by a flat-topped hill called "Level Mount." The place was already old
when the Song Dynasty poet Su Shi made a tour here with his father and
brother in 1059 AD. He composed a verse in which he said, "Level Mount
is well known in the world, and I am confident that the life here is
leisurely." From then on the hill has been called Mingshan (Well-Known
Mount).

Fengdu is historically called "ghost city" that is, the nether world.
The earliest origin must be traced to the history of the Ghost nationality
belonging to the Qing tribe in the ancient Ba and Shu Kingdoms. At that
time, human beings were still in the primitive stage, with little productive
force and the least developed science and culture. In the tenacious struggle
against nature, they could not scientifically explain natural phenomena
such as thunder, wind, rain, the sun, the moon, the stars, and birth, age,
illness and death. They mistook all of these for things dominated by gods,
which is the source of primitive religions.

According to the inscriptions on bones or tortoise shells, and on ancient bronze objects, the area of Sichuan was called Guifeng in the Shang
and Zhou dynasties. The Ba and Shu were two tribes of the Ghost nationality. In the period of the Eastern Zhou Dynasty, Fengdu was one of the
capitals of the Ba Kingdom. There were incessant contact and mutual infli-
tration of customs, politics, economics and ideology between the Ba and the
Shu. It is commonly believed that at that time the Emperor Tubo was born.
He became the first Emperor of the Ba-Shu or Ghost nationality. Tradition
says that he resided in Youdu. Today the two characters *"Yu"* and *"du"* are
still left in the decorated archway of the Son-of-Heaven Palace in Mingshan,
Fengdu. This tradition is said to be the earliest historical record of the city.

The other main legend about Fengdu concerns the "Two Celestials."
According to the tales written by Ge Hong of the Eastern Jin Dynasty, Yin
Chang Sheng, a relative of Queen Ying, flew to paradise in 122 AD, from
Fengdu and thus became one of the celestials. The other was Wang Feng-
ping, an official who had cultivated himself according the Daoist doctrine.
He became a celestial in 233 AD. These two celestials were called the
"Kings of Yin" in the Tang Dynasty and mistaken as Kings of the Nether
World (*"yin"* in Chinese means Nether World). Fengdu has thus been
named "Ghost City." The city has another relation with Daoism. In 198 AD
Zhang Lu, the grandson of the famous Dao master Zhang Daoling, established one of the holy cities of Daoism in Fengdu, naming the Level Mount
one of the 72 holy places of Daoism. Since then Fengdu has been one of the
missionary centres of Daoism.

The Ghost City is also intimately related with Buddhism. The first
Buddhist Temple was built in Mingshan in the Eastern Jin Dynasty. The

two religions were often involved in political struggles, with Daoism being made use of by rebellious peasants. But Daoism gradually lost its position and was pushed out by Buddhism. The older Daoist buildings, the Palace of the Nether World Monarchs (temple of the celestials), the Ghost Gate and others were taken over and given Buddist legends.

Tiny Mingshan Mount has undergone a long process, from primitive religions representing witchcraft through the legends of Daoism and Buddhism, and today it is a popular attraction for Chinese people. It may have little of its religious significance left, but it is a tourist attraction.

Thousands Of Graves

(Ruth, Badong, October 31, 1992) We had reached the mouth of Shennongjia Stream and I looked for the graves that Mr. Tian Yuan, formerly of the Bureau of Cultural Relics, had promised to show me. I saw nothing but orange trees, houses, and gardens. But underneath, he said, there were thousands of simple graves, thousands of years old. They were all underground and unmarked, their exact locations known only to the Bureau of Relics. "We do not want to tip off grave robbers," he said. All would be flooded, their as yet unrevealed secrets never to be known unless disinterred soon. It would cost about ¥4 million.

The art or pseudo-science of *fengshui*, or Chinese geomancy is ancient, its origins probably three thousand years old. Any buildings, especially temples and graves, were carefully placed in harmony with nature — the cosmic breath — to ensure good fortune to the living. The Chinese believed in an afterlife and the spirits of the departed had to be kept happy because they influenced the fortunes of the living.

The *fengshui* of the graves here looked good, with running water in front and a mountain, the dragon, the source of good luck, behind. The mountain also afforded protection from the cold of the north. Many of the rules of *fengshui* were related to common sense. *Fengshui* means wind and water. It was good for graves to be exposed to mild winds, fresh air, to cosmic breath. It might take months, years, to find another proper *fengshui* site. For graves, the birthdate and animal year of the deceased were consulted. The shape and condition of the sand and the soil were important. Some rules sound like outright superstition. Good fortune comes down from the mountain and enters one's house. If the windows are placed in a straight line so that good luck can go out a window on the other side, that is bad. Or if a door is hinged so that the good luck can go straight out, rather than be blocked to stay in the house, that too is bad. The screen in front of many doors in Chinese buildings is not just for privacy. It is to keep good luck in.

Fengshui has affected the decor of offices, even offices of British officers of the Hong Kong Police. People have rearranged their furniture for it, and even changed the direction faced by their front door; it is best to face south. A woman friend in Hong Kong was told that she had to change her sleeping room or block the window in her current bedroom — or have trouble with men. The *fengshui* business is alive and well in Canada, too, where immigrants from Hong Kong, fleeing the uncertainties of 1997, have invested in buildings. Toronto has a corps of *fengshui* experts.

Ancestor worship has also been practised in Canada. When I was growing up in eastern Ontario, we used to go to the local cemetery every spring to clean and decorate the graves of the Chinese who had died. My father would pour liquor on the graves and burn incense, and all of us would stand up and bow our heads three times on command, in respect. We would take a picnic that would include barbecued pork and rice, and eat around the graves, sometimes leaving a bowl of rice by the headstone for a while. We would be sharing the essence of the food with the departed. It was not a morbid occasion.

A few Chinese still do this on the *qingming* festival in China, especially in the south. But in Hong Kong and Singapore it is virtually a public holiday and thousands visit the cemeteries. But do the Chinese in China today care about *fengshui* and ancestor worship? Would the government consider it when relocating these old temples and tombs? Critics of the dam point to these ancestral graves but I do not think many Chinese people care about them now. If any do, it would probably be only the elderly. The young do not, unless influenced by the Overseas Chinese. But as Overseas Chinese invest money in their ancestral land, the practice is being revived.

As for good fortune, it has not really worked in Badong. The city is not wealthy. Maybe the graves are incorrectly sited. *Fengshui*, however, seems to have worked well in Hong Kong and Singapore!

Ancestral Graves: New Concepts, Old Traditions

(Kailin, Fengdu, September 17,1992) According to the traditional patriarchal clan system, the ancient family grave is a symbol of the ancestors and the continuing blood relationship. Respect and worship of ancestors are reflections of this tradition. The ancestor must not be insulted or offended. So moving the ancestral graves can be thought of as an act of destroying the clan and the family, an act of evil. In traditional thinking, this harms the souls of descendants and produces a lack of balance in their souls.

But in modern China a new concept has been challenging this old tradition. The need for modern facilities (building roads, extending factories), has made moving ancestral graves a common practice. A majority of contemporary Chinese people accept this. Although some of them are reluctant, they obey for the sake of public interest, without any hostility or insult. Only a small minority still use the method of earth burial; the majority accept cremation because of sanitation and the need to save land. Cremation is beneficial to society and to people, and so more and more people prefer it to burial in the earth.

According to the traditional Chinese idea, the ancestral grave should be in one's hometown. Originally my family's ancestral grave was beside the house of my grandfather in my hometown, Joujiang City, Jiangxi Province. Then my grandfather was transferred to the St. Johns University in Shanghai. My father followed my grandparents and settled down in Shanghai. During the War of Resistance Against Japan, all the members of my family wandered far from home. Later, after Liberation, many members of our family returned to Shanghai. Since my father met my mother in Chongqing, we began our new life in Chongqing. Therefore, nobody remained in our hometown, Joujiang.

In 1953, our Chinese government needed a piece of land, our ancestral grave, to build a museum. My grandpa agreed and presented our government at no charge with the land and the house. At the same time, he moved our ancestral grave from Joujiang to Hangzhou. I was told that the plot had a good *fengshui* (auspicious placement in the landscape) and was near Shanghai.

Every year many members of our family, except my younger sister and me, visit the grave to honour the memory of our ancestors. My oldest uncle, who had been living in the U.S., died on the way from America to Hong Kong. The box containing his ashes was sent there. So were my second uncle's and aunt's. But when my father was alive, he had a new, modern idea. He said to us: "I don't want to interrupt or disturb anyone. After I die, you won't play funeral music or light firecrackers." (Usually, very loud funeral music is played and firecrackers are set off for at least three days.)

When my father left us without saying a good-bye (a third sudden cerebral hemorrhage), we followed his will. My father was selfless and correct. Then my mother, sisters and I discussed whether to send the box of his ashes to Hangzhou. All of us said: "No, because it is far from Chongqing and not convenient to visit." So I put the box in a wonderful house, Ping Dingshan in the crematorium of Chongqing. Ping Dingshan is a place for the boxes of ashes of important persons. My mother, sisters and I think it is an ideal place for my father, because it wouldn't interfere with anyone.

I think my grandpa was correct to move our ancestral grave because the action was beneficial, although it sacrificed our own interests. I often tell my son not to be afraid of sacrificing his own interest. Every year, since my father died, I have taken my son to Ping Dingshan to honour the memory of our ancestors. Usually I ask my son to do something good, such as doing chores at home, to show respect and filial piety to his grandparents and maternal grandmother.

However, I don't want to criticize the traditional ideas which were formed over thousands of years. With the development of society, people distinguish for themselves what is right or wrong. I think things which happen naturally conform to the historical trend of the times.

"Drive Those Engineers Out!"

(Caroline, Shibaozhai, September 17, 1992) The rain is just beginning and the wind stirs small whitecaps on the broad expanse of river that we are approaching. Robert goes on deck. He has braved the North Atlantic on a destroyer in winter so this gale is nothing much! The door to the front deck bangs in the wind and he secures it with a chain that hangs from the push bar. The lights on the buoys along the channel come on (an innovation, we learn, only from 1985). We are very close to the south shore now and a ledge extends almost the whole way across the channel at one point. Small fishing and gravel boats have taken shelter behind the rocky outcrops.

The next stop, Zhongzhou, is called and we pack and go to the exit. But Li consults with a friendly policeman and it is decided we should go on to Shibaozhai and return here later tomorrow morning. We agree to pay ¥40 to the captain to stay in second-class splendour for another hour.

It is dusk when we arrive in Shibaozhai (or rather, at Xitouzhen across the river). We trudge up the long stairs, and Li starts asking anyone we meet where we can find a boat to cross the river. We are directed back down to the flat expanse of water-smoothed stone on the far side of the port where they fix boats. The creatures taking wing about us, Robert says, are bats. On and on we trudge. We have been up since 6:30 (5:00 for Li). Li is now only a small figure in the distance, in her brilliant green T-shirt and green and magenta backpack. I struggle to catch up with her, feeling the weight of my pack and packages, anxious not to let her down, almost stumbling over a wire cable. The others trail far behind.

I arrive at the water's edge just in time to spoil the negotiations Li has been carefully conducting. The deal was almost done, but now there is a new element: a foreigner. Who can take the responsibility for her safety? The boat is old . . . if anything happened there would be trouble. There are five or six people on the little dock, including one woman who takes the lead, and another who must be the local crazy. She laughs and leers, talks to herself, interjecting her merry bit into the brouhaha. Perhaps this is all for nothing as the captain of the boat is not here.

By now everyone has arrived and the damage is thoroughly done. Kailin tries to sniff out if anyone really is in charge. We talk money. ¥50 is agreed upon. The regular fare would be a few *fen* a person but clearly there is no regular service at this hour!

Finally the captain comes and leads us over a series of gangplanks and through beached boats to the small wooden tug that will make the trip. We are ushered down into the cabin, which has four crude bunks. We pile in the luggage and the ladies dutifully sit down, as instructed, and wait. The gentlemen go on deck. Robert is absolutely delighted. It is now black dark. The searchlight of a passing ferry sweeps the stone walls of the far bank and momentarily rests on men fishing from their sampan. The rain has stopped, the wind has died down and it is completely still: a magic night.

Too soon we clamber off in the darkness, over more planks and up more stairs. We are met at the top by a local woman who leads us through the village and, eventually, to the local keeper of relics. We literally feel our way through a dark and narrow passageway no wider than two arm-lengths, touching one side then the other. Then the roadway widens out, but only a bit, into a street lit by the open storefronts. Each one has a single 20-watt bulb supplemented perhaps by a candle, perhaps by a black and white TV set, and so the effect is of golden candlelight and shadow and weird blue flashes. The image of a boy lingers, sitting doing his homework, at his family's small wooden dining table, under their one bulb, candles beside him. It is a tableau that could only have been painted by Caravaggio. Families are cooking supper and watching the same program on TV, and we briefly look through their doorways and storefronts and into their lives. No one pays us any attention.

Then, ahead, we see the very bright light that is our destination. The Hot Pot Restaurant and Karaoke Bar is the "sideline occupation" of the new Keeper of the Relics. Li has come armed with an introduction to the previous keeper, who has a good reputation, but we will have to deal with his successor, who is an unknown quantity. We are warmly welcomed, so unexpected and out of the night, ushered into the restaurant and offered tea. We refuse supper as we have eaten on the boat.

Although we pull out our mugs for the tea, Robert and I are too embarrassed not to sip a little from the china cups so kindly provided by the enthusiastic teenage waitress. When she is gone, we think better of it and pour most of the contents out the window, hoping that no one is below. Our friends feel no dangerous compunctions to politeness and their tea remains untouched. The new keeper of the relics will take us to Shibaozhai hostel once he has finished his dinner. Zhongling and Kailin soon dub him "The Promoter." We wait, but not for long. The path and stairways up to the hostel are also unlit. We stumble as best we can in the dark (my flashlight, soon to be lost, is mysteriously forgotten, buried in my pack). That pack is getting heavier and heavier and the stairs go on and on.

The temple-style guest house is new. It apparently has two bedrooms, four beds and four or so occupants. They have settled in for the night. Negotiations commence. That we will stay in some of those beds is not in doubt. What will these men do? Kailin shrugs. It is up to our host to, "Drive these engineers out." They are here for a conference. I protest that this is unfair and get a contemptuous look from Kailin. "This is China." Money talks, but

I am not sure to whom. ¥20 is produced (a small sum) and the engineers, without any discernible rancor, pack up and leave.

The Promoter agrees that for ¥150 he will pick us up at 7:00 a.m. in his van and show us the famous local Han tombs, the *Dingfan Two Que* (sic), which is in the CYJV Report, and something else. We will breakfast in Zhongxian County town, the town we passed by yesterday, and return to see the temple at Shibaozhai, catching the boat for Wanxian in the early afternoon. Sounds fine.

The pavilion is surrounded by pine trees and there is a fresh breeze (but no water for washing, we learn). It has an outhouse set on the edge of the cliff, and a light bulb with a string pull, to guide our way. There is lots of ventilation, the droppings drop down the hill. Clearly it is not a temple, and it is not old. Nevertheless we are delighted. The men get one room and we take the other. Ours has only two beds so the two smallest, Kailin and Li, take the biggest one and will snuggle in head to foot. There are the traditional rope springs, a cotton bat mattress, sheets and a quilt. Just fine.

First, however, we have a meeting in our room. Zhongling climbs the ladder to investigate the trap door to the roof and discovers an enormous spider on the white-washed wall. Li grabs a twig broom and scoots up the ladder after it. The spider is too fast for her furious blows. Everyone laughs. Li must have been watching too many Gong Fu movies this afternoon. The spider skitters safely down the wall behind the door, only to be stomped to death on the floor despite our half-hearted pleas that spiders eat mosquitoes (there are a lot of them). Robert and I try to get Li to talk about the site we saw this morning but it is hopeless. Our translators refuse to translate and our expert thinks work is over for the day. I try to clarify that the famous drum is bronze, not copper. Kailin insists: copper. Li says that actually very little is known yet about the Ba, and that is that. These may be Ba graves but that does not mean that the Ba made the grave goods. She promises an article, but we protest that then we will have to find someone to translate it. Tough. It is 10:00 p.m. As expedition leader, I declare the meeting over and we all fall into bed.

Who cares about the bellicose Ba? There is not, to my knowledge, a Ba people in modern China calling for their rights. Sichuan has seen so many migrations and plagues and wars that the legacy of the Ba and the Shu is under ground. But there is, still to solve, the mystery of the origins of Chinese civilization and language. And the Ba themselves. Did their big noses come into China from the far west? or from the south?

4

SHIBAOZHAI AND ZHONGXIAN

The Many-Storeyed Pagoda

(Robert, Shibaozhai, September 17, 1992) The tugboat labours against the current. I can smell gas fumes, rope and the dankness of the water. The tug's searchlight probes the darkness of the river ahead. From my perch, no longer on the upper deck of a passenger boat but on a much smaller craft just an arm's reach from the surface of the water, I can appreciate what a busy road the Yangzi is even in the dead of night. Suddenly a small boat is caught in the beam of the lamp. A fisherman standing in the bow hurls his net into the swirling water. The tug manoeuvres to miss the smaller craft and again there is nothing around us but darkness. It strikes me that the whole trip is much like this moment. Foreigners can only glimpse fragments of the reality, even in the light of day.

The wind is warm. The boat's searchlight illuminates some of the ridges of bed rock that thrust up everywhere out of the river. Again the bluff-bowed work boat manoeuvres. Now, in the night ahead of us looms the dark mass of Shibaozhai, or Precious Stone Castle. In the dusk, when our boat passed by before landing on the far bank, we saw the solitary block of stone that towers 60m above the riverside village. But from close to water and in the faint light from the village nestled at its base the rock looks even more impressive. Against the night sky I can also make out the up-turned eaves of the pagoda that climbs the side of the rock and then reaches above it, a dozen storeys in all.

I am immediately and totally captivated by this place. Our boat pulls alongside a floating wharf and we scramble ashore over the precarious network of catwalks. The village is in almost total darkness. The electrical supply to rural areas is even more meagre than in the cities. We trudge up the stone ramp and through a passageway that is darker still. As we emerge into a narrow street I have an overwhelming sense of time travel. I feel as though I am transported back through the centuries. The feeling began yesterday when I saw men working in smoking brick and lime kilns. These are industrial processes that I only know from books and archaeological remains. We had also seen blacksmiths forging farm implements and parts for boats. They work in ways that are only used nowadays for demonstration purposes in Canada. But as we walk along that winding street with its two-storey, half-timbered, overhanging houses, I am in the 19th century.

It is still warm and, while the shops are closed, many of their doors are open. Each revealed room has only one naked bulb suspended from the ceiling. These lights, however, are of such low wattage that the effect is more like candles. In one doorway, in fact, there is a candle, sitting on a table flickering over a small boy doing his homework. By about age ten Chinese children need to have learned at least 500 characters (out of over 20,000).

We were being led through the village by a person Li has met near the boat dock and has asked for directions. We are looking for our "connection" in this particular place. I remain quite unaware of the difficult situation that is unfolding until well into the next day. For now I am soaking up the experience. After sitting down in a very comfortable restaurant and having some tea, we are ushered back into the street and begin another circuitous walk through laneways and passages. One doorway leads to what I think is the local museum. It turns out to be the town's karaoke bar. My time-warp is taking on surrealistic overtones. Finally we begin to ascend what seems like an endless stairway. The stone steps are not enclosed but in the darkness I can see nothing. We go up and up. I get more and more excited. When we emerge we are at the local Bureau of Relics Guest house far above the town. The doorways are round, the shutters on the windows made of wooden lattice-work and the tiled eaves turned up at the corners. I think we are in the very temple at the top of Shibaozhai.

Caroline is obviously more aware of a couple of impending problems than I am and she is not sharing my ebullience.

"Don't get too excited," she says. "It's not a real temple you know."

"Oh, it may not be," I reply, "but it's a real tourist attraction."

Shibaozhai is like Chongqing and the river boats in that it is not, close up and in the cold light of day, what it appears to be from a distance and in the dark. The guest house is not the temple at the top of the massive stone block I had seen from the river. It is a sort of pavillion on an outcrop or ledge of rock about a quarter of the way up. Later the next day, when we come back to the village, we will also see just how much of a real tourist attraction Shibaozhai is.

It is probably inevitable that a place of such physical distinction and beauty would have become a shrine and a seat of legends. One figure to whom the site is dedicated was a local Ming Dynasty heroine named Qin Liangyu. In the ongoing and probably futile attempt to correlate Eastern and Western legends, she is sometimes referred to as China's Joan of Arc.

But the image of a young woman warrior is about all Joan and Liangyu have in common. Liangyu's husband was a general during the time of Emperor Wanli (1573-1620). When the general fell in battle, Liangyu took over the leadership of the troops and defeated a rebel army, delivering the towns along the Yangzi. She is represented at Shibaozhai, fierce-faced and in full armour, astride a white stallion. Fu Kailin is anxious to have her picture taken in front of the statue of this warrior woman.

Another legend attached to the place concerns the miraculous supply of rice. It seems that during a time when Shibaozhai was a monastery a priest put a handful of rice into a hole at the summit of the rock. All the monks could fill their bowls from this source, eating as much as they wanted, but the supply never failed. This miraculous situation continued for a long time, but finally sometime around 1800 AD, rather late for legendary events, a certain greedy monk broke the spell. He tried to enlarge the hole to increase the yield of holy rice so that he could sell it. The offended gods, of course, cut off the supply and although the hole remains, the rice and the monks are gone.

We had the opportunity of visiting a couple of Buddhist Temples in Chongqing. They were functioning temples in the sense that there was a resident religious community (or a priest at the very least) and their main purpose was devotional. Shibaozhai had certainly been such a place in the past, but now it is a state-run historic site, a tourist stopover.

The structure itself is somewhat of an enigma to me. Some sources claim that the temple is 1,500 years old. While I have no doubt that there was a shrine on the site that long ago, it was most definitely not the one we see today. The present building is said to be hundreds of years old. The form may go back that far, but how many times have various timbers been replaced? Has it been re-roofed? Has it been totally or partly rebuilt?

I am looking as closely as I can at Chinese buildings in general and ones like this one in particular. I came to China to get some idea of the likely effects of massive permanent flooding on the culture of the region. Insofar as I have any claim at all to expertise, it is as a social historian and analyst of built heritage. But what I am seeing is not anything like European or North American building. By that I don't mean in a stylistic sense with reference to particular roof lines or decorative detail. What I am beginning to understand is that the Chinese have a totally and fundamentally different attitude and approach to the concept of what a building is.

The buildings I have seen so far — houses, shops, workplaces — seem to be of three basic types. There are what appear to be traditional and old-looking one- or two-storey structures. As the trip goes on I will come to understand a bit better how these traditional structures are put together, as well as how they come apart. The second type of buildings are brick or reinforced concrete. Often these have flat roofs. They are usually two storeys in the countryside, four- or five-storey low-rises in towns and grow into multi-storey high-rises in the cities. These concrete low-rise and high-rise buildings are examples of what we can call the "post-Corbusier," or "international style." The third type, transplanted European-design structures, are relatively rare except in places such as Chongqing, where there are some old banks, consulates and military buildings.

It is pretty clear that all the grey and repetitious concrete boxes have been built in the last thirty years. In spite of their lack of aesthetics these buildings are a tremendous accomplishment for a country like China. The few European-style buildings are obviously older. But it is a surprise to me to learn that the common, traditional-looking buildings that I see everywhere are also seldom more than a generation old. Zhongling's house, where I stayed in Chongqing, seemed to me to be at least a century old — it had been built after the Second World War. When I ask about the age of rural peasant houses, most often the answer was thirty, maybe forty years.

So here I am, a person completely out of my element. After all, I live in a 140-year-old Ontario house. I am used to an urban and rural landscape in eastern North America where a significant portion of the built environment is over one hundred years of age, and still serviceable, and where two-hundred-year-old buildings are found. I have studied many structures in Europe that have stood for five hundred, a thousand or even two thousand years. China to me is a place of ancient culture. And so it is, except that, unlike my culture, very little of China's cultural capital has been invested in permanent building.

There are no doubt many good reasons for this fact. The climate in southwest China is quite benign compared to the north European origins and the North American home of my cultural tradition. There is less need for stout, weather-resistant structures. At the same time the damp monsoon seasons have probably been harder on timber and brick than they have been on the people. For these and other causes the Sichuanese seem to have adopted a kind of disposable architectural ethic. Neither time nor skill seem to have been lavished on the permanence of construction. I watched some carpenters working in one place and was amazed at their rough-and-ready way of fitting and shaping wooden structural members. There are, of course, many wonderful buildings in China. The many-storeyed pagoda at Shibaozhai and the Zhang Fei Temple are among them. My comments about Chinese approaches to building are not criticisms, but simply observations of a different culture. It is my own notions of heritage conservation that have to be reconsidered in the glare of the Sichuan sun and the reality of Chinese traditions. That is what I am struggling with.

Bad Guanxi

(Caroline, Shibaozhai, September 18, 1992) Compared to yesterday, today is a disaster. The bad side of *guanxi* raises its ugly head. At 5:30 a.m. a trumpet reveille on the loudspeaker system is followed by insistent and aggressive music. Wake up time. I thought this violation of human rights had ended with the Cultural Revolution! We lie abed, enduring. Li, however, is up and bustling about. There are loud cries from down below our pinnacle. Li disappears: Kailin says the driver has come early and we must go. No way, I say. We agreed on 7:00 and even that is too early because at seven it is too dark to take pictures.

Zhongling and Robert pull their quilts over their heads and go back to sleep. By 6:45, however, Robert is up and delightedly using the tripod to try to take pictures of his beloved "temple" because it is still quite dark. We relent, but we ask, remembering yesterday, why not eat breakfast now? Our host is indignant: Zhongzhou is the capital!

Our chariot is a red six-seater Jeep Cherokee, probably with four-wheel drive. I climb into the shotgun position and we take off flying. While it is true that our friend has not had any warning of our arrival, and that in his mind he is doing us a big favour (for half a month's wages), his aim, we discover bit by bit, is to get our expedition over with as soon as possible because he has an important meeting back in Shibaozhai.

The countryside here is more prosperous than the Wu River Valley: there is much more arable land. The road to Zhong is only a bit wider than yesterday's but, clearly, it has been open longer. There is stone cutting, but less heavy industry. We stop. Can we walk several kilometres? Of course. (It is, perhaps, a kilometre in total.) Our host sets a very brisk pace across the paddy fields. Our destination is the "mysterious" Han pillar, Ding Fang Dui Que, mentioned in the CYJV report, at 158 m above sea level. It is to be moved to a museum, at a cost of ¥50,000 (US$12,000) because this beautiful valley will be flooded.

The lone stone pillar stands high on a gravelly but green slope above the flood plain of the Ganjin River. Stone blocks are scattered about and there are rectangular depressions around the pillar, where grass does not grow, indicating perhaps the remains of a floor. The river, which we will cross and recross several times, is very low.

Robert and I set about photographing the pillar. It is about six metres tall, with two hats: two roofs. Four small gargoyles, like caryatids, hold up each corner below the "hats" and vestiges of a dragon in low relief can be seen on one side of the shaft. No one, we are told, can make much sense of it. Usually there are inscriptions, but this pillar has none. Usually Han tombs had two pillars, now there seems to be only this one. There is no indication of a tomb or a local Han village, as best we can tell from cross-examining our host. What does the name mean? No meaning, it is just a name. Later Kailin gets a drawing of the pillar and says she can tell us what the badly-weathered carvings signify.

Then rush, rush. On into Zhongzhou. The promised breakfast at our guide's friend's restaurant is good; in fact it is the best breakfast we will have. By now we are matter-of-fact about our plastic boxes. The first *baozi* (steamed bread with pork inside) are cold, but the eggs peel easily and are fresh and warm, not green. When warm *baozi* are brought, and warm milk, we are content.

Then rush, rush. Through narrow streets and into a store and down a dark stair into a basement that contains a cave temple. The practice of building cave temples began in India and spread to China in the fifth century. We can just barely make out three grottoes, each with an almost free-standing carving of an enthroned Buddha. His elaborately haloed Bodhisattvas and arhats form a lively frieze behind him, alas with their faces gouged out. They are carved into the rock in a hierarchy of classes, with no concessions to perspective, the first row in half-round, the upper rows in low relief. Their hands are clasped in Buddhist salutation above their ample bellies and their robes hang, like angel's robes, in long folds. At the arched entrances to each grotto, and at the foot of each Buddha's throne, are less conventional creatures, human and otherwise, musicians perhaps, carved in the round, grinning, breaking free of their stone prison. The tableau is similar to medieval paintings of Christ enthroned in heaven, and it is common nowadays to look for borrowings between Christian, Hellenist and Buddhist art.

Li looks a bit green. The tourist brochure says that the carvings are from the Tang. Apparently she has not agreed with much our guide has said, but is too polite to contradict. The carvings are very weathered and in the darkness we can scarcely make them out. We do our best to photograph them. Either our friend does not know that there is a light or he knows that when you pull the string, it doesn't work. These sculptures, we are led to understand, will be saved, although they are not mentioned in the CYJV report. In fact they are Ming. The Tang carvings are elsewhere but we don't know that.

One of the precepts of Mahayana Buddhism is the Seven Dharma Methods, ways of accumulating merit and good Karma: building places of worship; gardens and ponds to provide cool shade; toilets and places of convenience; boats to help people cross rivers; bridges; and digging wells and providing medicine. Providing boats is especially useful in the Gorges but, says one commentator: "Since 'making images' could erase their cruel crimes and guilt, it is clear why the ruling class strove to build and repair temples and images" (Shi 1980).

In the street again, I suggest we buy a flashlight the better to see the Han Tombs. Our interpreters take this to be "flashcube" and we are off again running, block after block, looking for a camera store. Run, awkward foreigners run, with bags and cameras flapping! I am puzzled because we pass many stores that sell this common household item. We arrive at the camera store. Flash? No! Flashlight. We have wasted ten or fifteen minutes. Our guide is now letting it be known that he has an important meeting back in Shibaozhai. Again, we drive so fast we can hardly focus our eyes let alone our cameras and heaven forbid we ask to stop!

What about the Han tombs? We would have to climb three, maybe six kilometres to the top of a mountain and there is not enough time. I protest, "We paid, and we go." The boss doesn't budge. Our Chinese friends are embarrassed. You must see the pagoda at Shibaozhai, he says (that will take twenty minutes), have lunch, *xiaoxi*, and be on the boat at 2:00 p.m. "No rest," we say, "we're not here to rest." After our really good day yesterday, we are disappointed.

The jeep screams into the courtyard of Shibaozhai Pagoda and we pile our luggage into a corner of the gift shop. The open sore on Kailin's foot (no proper shoes and no water to wash it last night) is much worse: she says she will guard everything while we climb to the top. She visited here last spring with her mother. The foot is cracked and dirty but there is nothing to be done, short of buying new shoes or taking time to go to a doctor. Kailin refuses both.

The driver must be paid. I have decided ¥100 is enough. Our Chinese friends have laughingly dubbed our host "the Promoter" but they are too sadly aware of the futility of opposing exploitation by officials to stand up to him. But I am angry. We cannot afford, on our tight schedule, to lose a precious day. Perhaps if we push him, he will take us after lunch. Our little group is falling apart. Translation has stopped, concentration has lapsed, everyone is embarrassed and cranky. I grab Zhongling and tell him that he is to be the translator, not the negotiator, and to let me be the ugly foreigner. I then tell Mr. Promoter that we are angry, that we did not think much of the Ming carvings, which we had not asked to see, and that we had hired his car to see the tombs. Mr. Promoter does not fight much (he is due at his meeting). I give him the ¥100 and we set off to explore Shibaozhai.

Our gang is pleased: the enemy has been vanquished (nothing a Chinese hates more than being overcharged) and, better, in the confusion we have avoided paying the pagoda entrance fee. We are ahead of the crowds and have the place almost to ourselves. We had better get going, I say, if we are going to make the boat. (We do not have any boat tickets, but that is another problem.) My video camera mysteriously stops working, after taking pictures of the yellow dragons and the enigmatic figure beckoning from the upturning eaves of the gate.

Zhongxian County boasts a history dating back to the Neolithic. After Zigui County, our report says, Zhongxian has the most archaeological and historic sites. Legend says that the famous Stone Treasure or Jade Chop (Seal) Hill, is a precious gem left by the goddess Nuwa, the snail-maid, the ancestress of Mankind, and the goddess of matchmakers. The rebel prince Kung Kung broke one of the columns that hold up the sky, and its stump pierced the heavens. The corners of the earth gave way and violent rain and wind poured through the enormous black hole. Nuwa melted stones of the five sacred colours to repair the sky. One stone remained, the Stone Treasure that is Shibaozhai. A huge modern mural tells the story.

Stone Treasure Pagoda, so well known from its pictures, was built by? Well, we don't know. There are as many dates as guide books, and we have as yet to track down a scholarly article on this or any other historic building. Probably they are all the dates of reconstructions on the site. To choose the earliest, it was built in the reign of Emperor Wanli (1573-1620). To cite the latest, the top three storeys, called the Kiuxing Pavilion, were added in 1956. This seems reasonable because Isabella Bird says it had nine stories and early photographs do not show them. These last three roofs, which are offset on the top of the rock, make the structure visible from all directions.

This famous landmark is not discussed in Chinese architecture books because pagodas are defined as stone or brick towers, built to house sacred scriptures. The Stone Treasure is an elaborate stairway to the ancient temple on top. "In ancient times," at one particularly steep place, a chain railing was installed so pilgrims could hoist themselves, hand over hand, up the crude steps hacked into the side of the cliff. An inscription at the gate proclaims: "Whomsoever climbs here must feel himself humble," "Only the humble may climb?" or, perhaps, "This climb will make you humble!"

The wonderful pavilion is all of red-painted wood, 56 m high, built on and dug into the cliff. Ladders take you up the 12 storeys. Each level has grottoes and resting places, and magnificent views of the river from the immense round windows. One pavilion, described as Two Dragons Play the Pearl, is hung with a traditional silk lantern and flanked by two cheerful yellow dragons who offer a red ball (pearl) in their outstretched claw. No one seems to know the story exactly but they think it symbolizes protection from flooding.

There is a 360-degree view from the top. To the south is the magnificent river, bustling with commerce. Behind us, and all around as far as we can see, are immaculately tended fields. When the waters rise Shibaozhai will be an island. The citadel will remain, ten metres or so shorter, but the farmsteads that it has protected so long will be gone. If the pagoda is the "Qing Dynasty building" mentioned in the CYJV report, it is at 175 m and at the time of writing, no "mitigation" was planned, not even for the reconstruction of tourist infrastructure.

Jade Chop Hill is fortified with a wall referred to as Tianzi Temple. The more ancient Lanruo Temple, within the walls, is said to have been built in the reign of Emperor Qianlong (1736-1796) but, as usual, looks much older. The sketch map of the site indicates a wall made of bricks from the Han Dynasty. A tourist pamphlet mentions that a peasant rebel, Tanhong, camped on the top with his army, and constructed a fortified village.

I find Li deep in conversation with a young man, who is introduced to us as Zhen Xianlong. (I assume that their meeting is purely a matter of chance, our luck is improving!) They suggest that we return to Zhongzhou: Zhen will take us to the Han tombs and other sites tomorrow. We can skip Wanxian to make up the day. Kailin's foot needs tending and the promise of a good hotel and a bath in Wanxian are no little part of my hesitation. Li then suggests that we go on ahead and she will catch up with us. I like that idea even less. "We must discuss it with Robert and the others," I say, stalling for time. I am tired and hungry, and I am getting the names of all these places jumbled up.

Robert says we should go back. How? I suggest we rent a van or *kangfulai*. Li says our enemy controls the only rentable conveyance in the area: we will have to take a local bus. The bottom line is that we have no boat tickets for Wanxian and so we might as well relax and finish seeing the temple. Then we will see what comes up. My 35mm camera, whose light metre has been giving unusual readings, suddenly whirs and the film automatically rewinds.

China's Joan of Arc, Qin Liangyu (1616-1636), is depicted in full battle dress, on a magnificent white horse, red banner flying, leading her troops into battle. Her birthplace Qinjiaba is nearby. In the temple are several historical dioramas and a great wonder, the rice hole. Somehow, I am not in the mood.

But there are also ceramic sculptures of Ba Manzi. Ba Manzi was a famous Ba general in the Warring States Period (Eastern Zhou). Civil war broke out in the Ba Kingdom and Ba Manzi asked the Chu State for help, offering three cities in return. Chu sent help and, when peace returned, demanded payment. Ba Manzi faced a dilemma: he had given his word but he could not bring himself forfeit part of his country. He offered himself in place of the cities, cut off his own head and had it sent to the King of Chu.

The King of Chu was touched by his suicide, relented, and buried the head in Chu. Since that time, in his honour, both the town and the county, have been known as Zhongzhou and Zhongxian, Loyalty Town and County. Down in the town, below the wall, is the narrow passageway where Ba Manzi is said to have decapitated himself. His tomb, however, is in downtown Chongqing, hidden by an overpass, enclosed with a grill-work and covered with grime. It took me most of a Saturday afternoon, with a student, to find it in 1990.

Lanruo Temple is a local museum. Some of the pottery figures from the Han tombs we had hoped to see at Tujing this morning are on display here, lively depictions in clay of the daily lives of the tomb's occupants, servants and companions, so that they might take their earthly pleasures to the next world.

It is, in the end, unsatisfying. The garish ceramic sculptures that Kailin regards as art, look too much like figures from a wax museum (admittedly wax museums are popular in the "West"). History turned into pop culture is not what I had been hoping for. The problem, besides our hunger and fatigue, is that they tell stories known to every Sichuan school child, but not to the foreign tourist. However, I am not alone in my ignorance. The common accounts of Nuwa do not mention Shibaozhai. The legend has been adapted to explain the local landform. And none of my Chinese friends in Toronto have heard of Ba Manzi or Qin Liangyu. China has such a wealth of myths and so much history that it is impossible to master it all.

We reassemble at the gift shop. The manager is anxious to make a sale. He has two "antique" padlock *qilins*, like the ones I have been reading about. In the old days anxious parents would hang pendants of this mythical unicorn around the necks of their children to protect them when they went outside to play. Kailin is suddenly in her element. She bargains him down from ¥100 to ¥70, still a tremendous sum for there is no silver in these chains. So *qilin* is our *quid pro quo*, and we could use some luck. The shopkeeper has some other old things and, as we are about to find out, that is because Shibaozhai is a major tourist town.

We are almost mowed down by a great crowd of people surging through the gate. A boat has just arrived (probably the 2:00 p.m. boat we were supposed to take). So Robert will taste tourism, Chinese style. The steep roadway (a passageway and stairs) is lined with stalls hawking tourist stuff. Kailin negotiates another *qilin* for ¥70, a much older-looking one. The local specialty is very finely-woven straw goods: round lidded baskets like porcupine-quill boxes, and slippers. Kailin bargains for a pair of old-fashioned hemp sandals with leather trim, the strongest and best ones I have seen. We develop a system. I float by and see what I like, describe it to her, and she goes back to buy it at the "Chinese" price. This stratagem fools no one, but is more fun than the temple.

By now, Li is far ahead, asking for the place to catch the local boat she has been told can take us back to Zhongxian. We are promised lunch on the boat. Robert is sure she is going the wrong way, but we follow her halfheartedly, drawn by the souvenir-sellers and the shuttered and barred wooden storefronts of the old Sichuan streetscape, with the overhanging, lattice-windowed living quarters above. The streets are paved with well-worn tightly-fitted cobbles. Like all tourist towns the deserted streets spring into abundant life when the tour boats dock. The most charming shop is kept by an old man and his granddaughter, who make paper mache dancing death masks of Monkey Kings and Buddhas and black-faced devils. I want one, but this time Kailin puts her foot down. It will just get broken, even the little frog. I settle for a photograph.

Finally, we are beyond the food stalls and the sidewalk restaurants. The streets are almost deserted, the proprietors having retired for *xiaoxi*. Li gives up on the boat dock but we have happened onto the "long distance bus station": actually, the point where busses pick up passengers for

Zhongzhou. There is an awful rusty old bus, crammed and ready to leave. I stall. We have not yet found a toilet or lunch. The bus pulls out. Surely there will be another.

Kailin finds a good toilet behind a construction site. We are afraid to leave to look for a restaurant and, besides, it is nap time now. We comb the local convenience stores (mostly liquor and cigarettes) for bottled water, buns, peanuts, whatever. The best we can do is a large plastic bottle of orangeade, and so we rummage in our bags for yesterday's white bread buns, a few peanuts and Kailin's earth melons. Robert has now decided that they are water chestnuts. The ones we buy in cans are conveniently cut up into small slices for stir frying. Well the texture is right and, perhaps the taste, but Kailin says NO. Out come the *China Daily's* again and we are picnicking on the sidewalk beside our mound of luggage, just as we did on the tail of the ship, but with far more meagre fare.

My mood has greatly improved by the time our keeper-of-the-relics comes by. Yes, obviously, we have missed the boat. No, we do not want to come to his restaurant for lunch. However, Li does ask to go up to the pavilion to look for her lost rice bowl. He has no hard feelings but we are not quite so generous.

A new minibus pulls up and stops. We pile our luggage onto those wonderful seats, to save them. Zhongling has negotiated a thermos of hot water from a local housewife for 50 fen and we make instant coffee in our mugs. I despairingly read the video manual, but with little hope of success. Li disappears to look for her bowl (worth a few cents, replaceable any-where). I start worrying that the *kangfulai* will leave: people are starting to get on. Kailin has bargained for a fresh but unripe fig; it is not so popular as the earth melon.

Li returns, none too soon, and we clamber onto the *kangfulai*. Kailin, whose foot is worse, dozes. So does Zhongling. People here are quite-sensi-bly conditioned to need their nap. Robert and I get a chance to talk. The ride, over the same road we had taken this morning, is spectacular. We are getting some good photographs of local industry, but we are not talking to enough people. When we are on the go, we have to meet the deadline of a boat or whatever, and at night we have been holed up in a hotel. How can we talk to people and yet avoid politics or attracting attention that might get our friends into trouble?

Robert is enjoying getting to know the road and we are seeing what is back from the river. I would very much like to visit one of the clusters of farm houses we see down in the valley. They are what I imagine a landlord's house to have been. Foreigners find China's reactionary past ever so much more romantic than the present! We agree that we need a meeting of the whole group, but only after we are washed, rested and fed.

The magnificent bus ride comes to its inevitable end. We pile out and as usual Li goes on ahead. The hotel Zhen has suggested has no lobby and, it seems, no name. The woman at the desk at the bottom of a staircase says it is better than the hotel at Fuling, but no, we cannot see the room first. Take it or leave it: this is the only hotel in town designated for overseas Chinese, Hong Kongers and foreigners. Foreigners do not like to be relieved of the burden of choice, even when it is for their own good and convenience! We pull out our passports and Zhongling starts the lengthy paperwork. The sun has finally come out and we are covered with two days' dust and sweat. A crowd gathers around the hotel door, gawking. We are like animals in a zoo, Zhongling repeats. Robert and I say we don't mind (we stare back at them), but our friends point out that we don't know what is happening.

Two cheerful girls grab at our cameras and indicate they are to show us to our rooms. I am in no mood to relinquish the cameras, even if they are not working. No one offers to carry the packsacks. For my stubbornness, I am punished with a six-storey climb with my pack on my back, cameras and shopping bag.

Our two dingy suites are side by side. Again, the toilet seats are cracked off and standing against a wall. But worst of all, the water is not on. We have forgotten that in traditional Chinese hotels the water is often limited to twice a day. The very nice floor lady offers to carry up some buckets to last until they can get the water on at 8:00 p.m. I am beginning to think like a Chinese: "Fine," I say. My hands and face are filthy and better her than me. The buckets splash water everywhere and soon the floor and everything else is muddy with our dust and soaking wet.

Water for tea is brought to our sitting room. Mr. Zhen is expected at any moment. But first, Zhongling has been investigating and insists that we decide on the spot what class of boat ticket we want and when. Second class? Third? The doorman will get the tickets but he has to go now. Li says skip Wanxian and buy the tickets through to Yunyang. I say let's discuss it. Again, everyone talks at once. Zhongling insists we decide about the boat tickets. Robert asks when? Will we have time to see the promised sites?

But now, our Chinese friends have switched to something completely different and, it seems to us, that paranoia sweeps over them. They have not gotten the necessary "certificates" for the trip. Robert points out that no one has asked us for any identification or even what we were doing (except for the police at the way station on the road back to Fuling and that problem, if there was one, seemed quickly solved). Our presence in any town is pretty obvious. It is said darkly that this is a very backward area. Robert and I try to explain the meaning of "I would rather apologize than ask permission" but to no effect.

Zhongling has also been told by the helpful doorman (the hotel didn't seem to have a door!) that the WAIBAN (foreign office of the city) can arrange a car for us, hotel reservations, everything. He darkly mentions the Beijing big shots here to investigate, taking up all the hotel rooms. I point out that the WAIBAN will be fussy about permits and certificates (not to mention charging a lot and that we do not want to be limited by their schedules or constrained by their guides). Robert and I think that Zhongling is just a bit too anxious to go deluxe or to go official. Or perhaps he is taking our griping too much to heart. Naively, perhaps, we think we are doing just fine, this morning not withstanding.

These problems are resolved by the timely arrival of our Zhen Xianlong from, we now find out, the Office of Relics Management of Zhongxian. He is arranging transport (another *kangfulai*) and he is trying to get reservations at Wanxian's best hotel, the Taibai, which has been recommended by Chongqing CITS as the only place on this piece of river where anyone would want to stay. Not to worry.

I point out that it is six o'clock and we have not had much lunch (although Zhongling boasts he has had three lunches). We are inviting Zhen to join us, can he suggest a restaurant? Do I think this is Toronto? Zhongling again objects that we will be stared at like animals in a zoo. Kailin has a darker argument for staying here, but she cannot speak of it now. Later she tells us that if the police saw us, because of the crowd, they would do a song and dance, about how dangerous it is in this backward area and how difficult it is to guarantee a foreigner's safety, and demand protection money. Again, certificates are discussed. The consensus is that it is better to lay low in the hotel dining room.

A Countryman of Dr. Bethune

(Robert, Zhongxian County, September 18, 1992) The hotel in Zhongzhou is described as "the best one in town." As we proceed down the river we almost always end up each night in the best hotel in town. My suspicion in most places is that we are in the only hotel in town or at least the only one that will take foreigners. But I shouldn't complain. If the floors are clean, the roaches smaller, the beds made. . .if there is any hot water, any water at all for that matter. . . if there is an elevator, paint on the walls and if the bathroom has something more than a towel stretched across the long-departed window. . .if these and a few other things are the case then perhaps it would be a passable spot.

There are pretty flowered spittoons in the halls of the hotel in Zhongzhou. That is a sure sign of the classiness of the place. When I use the seatless toilet in our room I have to flush it by putting my hand into the reservoir and pulling up the ball-cock, which had become detached from its chain. Water gushes from some unsealed joint behind the bowl. Still, once having learned what to expect, these places are welcome refuges after our long and eventful days.

The best parts of stops are always the mealtimes. The importance of eating in Chinese culture can hardly be over-emphasized. What expresses it best for me is the realization that the characteristic and not at all unpleasant smell that permeates every city and town we visit is the odour of cooking oil and sesame, which is used for flavouring. Breakfasts are generally catch-as-catch-can, midday meals sometimes fairly grand and sometimes modest, but our evening dinners are almost always the full treatment. That means dining rooms with many round tables that each might seat a dozen. We are usually joined by one or two of our local contacts, so our parties of six or seven make suitable gatherings for that very favourite of all Sichuanese pastimes. There is always a good deal of friendly debate, reluctance and equivocation about ordering. Since Caroline and I pay for the meals and are therefore technically the hosts, our Chinese friends have some misgivings about ordering. Caroline often joins these discussions but it is all too new to me. Only once am I given the task. My choices are politely accepted, but I order far less than usual and I think my offering seems a bit bizarre to our friends.

Sichuan is considered to have one of the four great cuisines of China along with Beijing (featuring Peking Duck), Juiangzu (characterized by its fetish for eating almost live fish) and Guangdong (the Cantonese style most familiar in North America). The Chinese consider these four, and French, to be the five foremost food cultures in the world. Although I do not yet understand it I am told that eating in Sichuan, as in other parts of China, is governed by a complicated set of principles, rules and conventions. This is particularly the case in the matter of ordering and thereby giving shape to the meal. The process is heavily influenced by Daoist theory of yin and yang. Such things as balance between hot and cool, male and female, and certain tastes such as sweet and spicy govern the choices. While I observe the choosing often it remains a mystery to me. Besides, at every meal I enjoy in China over the period of a month, there is something new at every sitting.

In spite of the questionable cleanliness of the towns in general, the streets in particular and hotels in detail, the food is almost always wonderful. The disarray of the kitchens, which we often have to walk through to get to the dining rooms, gives way to beautifully arranged plates of stir-fries and magnificently presented whole fish. There are soups, vegetable dishes, chicken and always pork in many and varied forms. Occiasionally there are a little more exotic items such as rabbit, salted fish and mutton. All too often, for my taste, there are sea cucumbers, or sea slugs as I come to think of them. The fish, however, is the *pièce de résistance* of every meal.

But meals are not only, or even primarily, gastronomic events in China. They have a central social function as well. This is true of course in most societies around the world but in China, where relationships among people are overwhelmingly important, there is special emphasis on dining together. On the occasions where our little band of five eats alone we deepen and solidify our friendships. Often, however, our meals are opportunities to either get to know or to thank our local contacts. These contacts materialized in every community through the magic of the letters of introduction that Caroline's Chongqing contacts have provided.

I have been duly warned of the propensity that the Chinese can sometimes display toward alcoholic consumption and carousing. Fortunately we experience this only in its mildest and most congenial form. The beer in the Yangzi valley towns comes exclusively in glass litre bottles and contains only 2.9% alcohol, compared to 4.5% or more in most Canadian beers. It provides a pleasant and relatively cheap beverage since water of the expensive bottled variety and overly sweet soft drinks are the only alternatives. The ordering of a couple of bottles of *pijiu* (one of the only Chinese words I learn) becomes a standard forerunner to our meal ordering. When we are hosting our local contacts at these meals, frequent and repeated toasting is also standard. Zhongling, however, never drinks the beer. He says that he is afraid it will make him fat. Instead he guzzles Chinese cola, which probably has twice as many calories.

On one of the first occasions of these pleasant dining and drinking evenings, our new-found local friend, Mr. Zhen, lifts his glass toward me and toasts. The translation is immediate.

"It is my very great pleasure to drink with a countryman of Dr. Bethune!"

This pleases me a lot. Norman Bethune, the Canadian doctor who died while serving in the Chinese People's Liberation Army during World War II, is one of my heroes. I still have the picture of him that hung in my cabin when I was in the Navy.

Pillars That are Statues of Guardhouses!

(Caroline, Zhongzhou, September 18, 1992) The way to the "deluxe" dining room leads through a public dining room and the grotty kitchen. Again however, the maitre d' and the waiter are eager to please, we eventually figure out just what is available, and the food is excellent. I specify no *hua-jiao* in the chicken with peanuts, despite Kailin's warning that it won't be authentic. But she agrees, "Foreigners can't bear it." Robert and Zhen spar off with litre bottles of local beer and even I try it. Out come the yellow plastic boxes, but the beer goes into the wine glasses provided. Robert upholds Canada's honour and he learns the word *gambei*.

We launch into an incoherent discussion of the "Han Pillar" we have seen this morning. Even Li loosens up a bit in the good cheer. In 1972 country people had built a brick structure to protect the pillar, but in 1987 the roof collapsed knocking it over. Zhen helped right it, but now it stands unprotected. It is to be properly housed in a museum — when they get the museum that has been promised as part of the budget of the Three Gorges Project. The county pamphlet shows a pillar, with a protecting brick roof and identical to our pillar, except that its base has been replaced with a brick pedestal. Are there two pillars? Or is the base we saw a copy? Although badly worn, it looked all of a piece. The two, if there are two, seem to be a pair.

Two kinds of pillars are known, we are told: the oldest were memorials in front of palaces or temples; later the wealthy had them built for their tombs. But stone pillars usually were carved in pairs, a child and a mother. Not just anyone was allowed a pillar and there are no inscriptions on this one to indicate that the builder was, as required, above the third rank. What of the four-horned beast carved on it? In ancient times the heads of sacred animals were symbols of riches and guardians against evil, door guards (*Pu Sou*). The dragon traditionally looks east, the tiger west, the snake north, the phoenix to the south. You can tell directions by identifying the animals. What was the extent of Han penetration of the area? Our questions are good, Mr. Zhen says.

No academic articles are forthcoming, but Kailin sends a drawing and the entry from a Chinese encyclopedia. It has little to add beyond the counting and classifying that foreigners often find so pedantic. Twenty-five Han pillars are known in China, ten are famous. Of the 16 in Sichuan, four are famous. This, I take it, is one of them. There are two kinds: single roof and double roof. Pillars were used as notice posts, perhaps to remind people when to pay taxes or to proclaim laws, or as a place for an inscription at the tomb of an important person. Robert later finds in his architecture book that the Han pillars common in Sichuan are guardhouses (*que*). However, they are not really guardhouses, but statues of guardhouses, carved in bas relief with the (two pronged) capitals of wooden pillars (*dui*) that are part of the traditional wooden post and lintel architecture.

After supper Zhongling and I go out to phone Norman back in Chongqing. Zhongzhou was Norman's father and mother's first mission post. The keylady takes the number and calls the post office, which will call back to the sixth-floor phone. To my astonishment this actually works. Norman is not there. We leave the message that we are alive and well in Zhongzhou. Norman calls back. His leg is still knotted up and he can hardly hobble. Everyone suggests massage.

We reconvene in the "suite" and pepper Zhen with still more questions. "Ganjin," as it is known, is a "key point for scholars now." He draws a sketchmap of the sites he will take us to tomorrow. First we will see two Neolithic sites, one on the Yangzi and one on the Ganjin River. Then, the Han tombs. We will have lunch at the small town of Ruxi and say goodbye to Zhen there. The driver will take us on to Wanxian in the afternoon. Wonderful. We are all too tired to pursue the Neolithic further. Zhen departs, and the party breaks up, for writing, washing and to bed.

As promised, hot water comes on at 8:00. In scoots Li, with her laundry. When she finally emerges, the bathroom is awash and wet clothes drip everywhere. Zhongling is first into the men's bathroom. He locks himself in and has to be rescued. I am next but the hot water, alas, is finished. I step gingerly into the corner of the immense bathtub for a few cold splashes. A lesson: she who gets there first gets the wash. Show no mercy.

Much later, I am still awake, writing up our day. All of the taps suddenly gush water. Cold water. I cannot locate the slippers provided. I get up and wade through the damp to turn off the taps.

Drawing of Dingfangduique, the "mysterious Han pillar in Zhongxian County. "

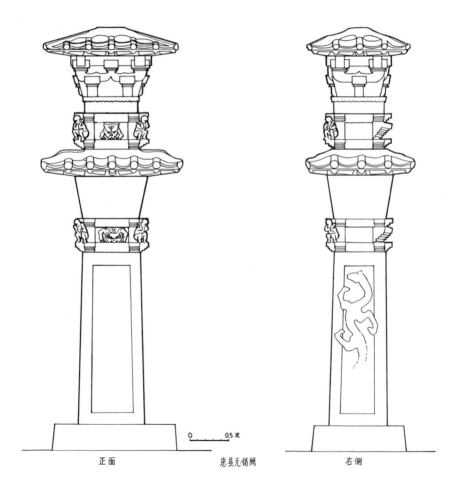

正面   忠县无铭阙   右侧

The Valley of the Mattress Makers

(Robert, Zhongxian County, September 19, 1992) Fortunately there are times when problems, such as the runaround we have been given by the local "operator," can end up having beneficial results. Our unique and detailed look at the Ganjing Valley and surrounding area is such an occasion. We end up going over some of the same terrain four times in the course of two days. The normal routine for tourists and even peripatetic researchers is to traverse a given area only once and, as that naturally dictates, in only one direction. Because we saw much of Zhongxian County north of the Yangzi a couple of times and from opposite viewpoints we came to have a more intimate picture of at least this one area that will be affected by the flooding of the big river.

We rushed from Shibaozhai to Zhongzhou early yesterday morning. The valleys are quite different from the Wu River, ascended the previous day south of Fuling. Instead of being steep-sided and rugged these areas are much more settled and hospitable. The scene is like the classic Chinese painting. The mist-shrouded mountains form a backdrop. The low green hills in the foreground are terraced all the way to the top. There are a few spindly trees either standing as lonely silhouettes on the hills or clustered in little groves in the clefts of the land. Here and there a thin silver line glinting in the morning light reveals a waterfall and marks the course of a stream running down the slope to join the river winding along the valley floor. Set throughout the paddy fields, in what seems like a random way, are little clusters of farm buildings.

When we pass by first thing in the morning there are peasants already trudging up the roads with heavy loads in baskets on their backs or mattocks borne on their shoulders. There are small children with backpacks hiking along in the gloomy dawn. Smoke from cooking fires wreaths almost every farmstead. There are road workers riding by in trucks filled with broken stone. In a large site along the road in one of the villages people are setting up moulds of some kind. At one turn in the road I catch a glimpse of a hill across the valley that has been cut in half as though it were a loaf of bread and some huge knife had sliced through it. On the crest of another hill I can make out a rectangular hole cut into the stone. It is large enough for person to stand in upright.

By the time we cross and re-cross these roads twice in each direction over two days the patterns of the countryside no longer appear haphazard to me. The answers to many of the questions raised by what I see have emerged, either from further observation or because I have had the time to ask Kailin or Zhongling, and they have either explained what they know themselves or have in turn asked others.

Over the area we cover in these two days I see only two schools. But they are a long way from some of the homes and that means very early starts for some of the children. It shows that getting an education must be a considerable struggle for them. I learn that the trees are sparse on top and stripped of their lower branches because twigs and brush are important cooking fuels in the countryside. I also see disks of dried animal dung hung on the walls of houses. I suppose that also to be a fuel supply.

The road crews dump crushed stone in piles along the right-of-way and drop off a worker with each pile. It takes them most of the day to distribute their load of stone along their section of the road. A certain number of these workers are women. The moulds in the industrial yard turn out to be for concrete slabs about three metres long, 60cm wide and perhaps 10cm thick. They are not solid but have several holes running along their length. I find out later what these are used for and why they are produced in this standard size. The process from mould laying to completion takes two days.

The third time we pass the hill that is sliced in half, I finally get the story. During the period beginning in the late 1950s, known as the Great Leap Forward, there had been a plan in this valley to divert the course of the river. This incredible cut, which looks more geological than human in origin, must have been the result of extraordinary effort. But the river still flows in its natural path. As for the straight-sided hole carved out of rock on the distant hill, no one had the slightest idea what it might be. Not all my questions are answered.

As we travel through these tributary valleys north of the Yangzi, Caroline is noting the archaeological sites, photographing potsherds, scrutinizing the strata in cut-banks and making systematic enquiries about the historic periods and minority peoples represented by the artifacts. I am still soaking up the atmosphere and trying to orient myself in rural China. The bus we have rented in Zhongzhou City stops by a long bridge leading into a modest-sized town. While I pause to photograph the carved stone lions that decorate the ends of the bridge, the others set off down into the dry riverbed in pursuit of our guide, from the local Bureau of Relics, Mr. Zhen. They are all far ahead of me. I suddenly become aware that beside the road, on the hillside flanking this town there are dozens or maybe a hundred people working.

I have begun to use my small dictaphone to take notes when I am on the move and unable to write. I click on the tiny machine and start to talk. By way of recording some of what I'm hearing here in this valley I'm going to try to get the sound of the hammers as I hear the workers breaking stones. It is an extraordinary rhythmic metallic clinking noise, without interruption from voice or machine. So right now I'm walking through a boulder field in the bottom of this river valley, not the Yangzi but its tributary, which I think is more or less called the "brine river" because there was salt found here from the earliest times. I see the shell of a small river crab lying on the rocks here, and also what appear to be snail-like shells. This reminds me how little animal life I've seen so far in China.

We work our way back up from the river bottom by way of an old stone stairway into the town itself. Between the buildings and the river we pass a large, round, stone structure that is said to be a reconstruction of one of the ancient brine wells. We find ourselves once again on one of those meandering medieval-like streets. Then I come upon a scene that leaves its image on my mind's eye more than most others I see on the trip. Two big wooden shutters are opened in the front of a shop and two men are working inside. I am told that they are traditional mattress makers. Apparently such mattresses, over rope springs, are the preferred bedding even in the cities.

Between the two men is a trestle table of some sort with a bat of cotton spread on top. It looks like there may be tacks or pegs around the edges of the table. One man, the older, holds a stick about a metre long, with a small notch cut in the end away from his hand. From a spool mounted on the wall a heavy thread or string passes through the notch in the stick. The younger man catches hold of this string when the other man pushes it toward him. Each time he takes it he appears to loop the string around one of a line of nails or pegs on the table around the edge of the mattress. The older man then takes hold of the string as he rocks back and pulls it towards him. In this rhythmic way they lay line after line of thread across the mattress first diagonally, then from end to end and side to side. This work is absolutely silent and has the beauty of a dance. The tension of the thread must be kept at perfect level all the time. Each movement of the older man has to be matched exactly by the younger.

If we assume for the moment that this is a father and son, and that the old man had once been the young man with his own father, and so on back in time, then truly this is the thread that runs through the ages in this remote valley in China. I think that this building and this village may be under several metres of water 15 years from now. These people will be. . .God knows where. There will still be thousands of mattress makers in rural China but in this place the thread of time will be broken forever.

Neolithic Sites and Han Tombs

(Caroline, Zhongxian County, September 19, 1992) At 7:00 a.m. we gather in the hotel dining room, the part for locals this time, for *baozi*, hot and fresh, with a large portion of pork inside. There is *xifan*, rice porridge, and hot water for instant coffee. Zhen Xianlong has arranged the *kangfulai* and the day is bright but not too hot. We pile everything into the bus. Robert and I sit together, the driver's grandpa beside us, Kailin is in front, Zhen and Li behind, and Zhongling is in back. Onto our road: ours because we have been over it three times yesterday. A few kilometres out of town, we turn off, follow along the Yangzi a short way and stop.

Last night Zhen had told us that this site, Ganjinggou, had been found in 1957-58 by the Yangzi River Power Authority. They made a brief investigation but were stopped by a tremendous drought and then a flood. In 1959-60 they arranged a house for the archaeologists and for farmers to assist them, but this time a fire stopped the work. Before the Cultural Revolution, if I have the names right, Yuang Yourun and Yuan Mingsen excavated here. In 1979, Sichuan University took over. In 1984-87, extensive excavations were finally undertaken. The CYJV report lists its elevation as 153m; it is below the NPL of the reservoir.

Li and Zhen, the professionals, have their well-sharpened trowels in hand. First we stop at a rammed earth house: up close we can see shards sticking out of it everywhere. We are delighted and then alarmed. We will see the same thing in other places: hopeless riches, a site churned by eons of farming. Thousands of shards, but shards from different periods, and nothing that can reliably be associated with them: bones, seeds, pollen, stone tools, nor the remains of hearths or dwellings. In archaeological jargon, these shards are "secondary depositions."

Archaeology proceeds by the simple and obvious rule that the deepest layer (all things like moles taken into consideration) is the oldest. Much of their most rewarding work is digging up prehistoric garbage dumps. Early people had annoying habits of keeping their houses swept clean and of taking their possessions with them when they moved. Mostly what archaeologists find is what went, broken, into the trash. Archaeological work involves painstakingly scraping down a site, centimetre by centimetre, distinguishing one layer from another and keeping track of what was in each, including a very careful analysis of the soil and what is in it.

This method can determine, for instance, that cultivated sorghum, appears at a site in all levels above (later than) level ten. Artifacts similar to those found with the sorghum may be found at another site and so the two sites may be cross dated and perhaps the spread of sorghum cultivation can be traced. Thanks to modern physics, many, many things, from charred seeds to chert and shards, can be given absolute dates within specific ranges of error. However, it is not simple and it is not cheap.

Every modern science, it seems, is being applied to the study of the past and data from new technologies is becoming available with bewildering speed. Because some are so new, their methodologies are untried and their results must be carefully evaluated. As in other fields, the scientist who gets the headlines in *Nature* or *Scientific American* (whose articles are then summarized in the popular press) has a better chance at the grants.

Archaeology has gone from being a rich man's obsession to a scientific free-for-all, but it is still mass entertainment. If the archaeologist is not a specialist in botany or geology or physics, he calls on those who are. The "-ologists" — the specialists in ancient flora, fauna, geology, metals, soil, you name it — are part of every modern excavation, and even more a part of the lab work that should follow. It is these specialists that China needs most, to train its archaeologists and technicians, to improve the quality of the work that is being done. There are many bright and enthusiastic young people, like Li and like those we met on the staffs of the local bureaus of relics, who are eager to learn and to make their own mark.

Zhen points to the vegetable patch behind us. He picks up one shard here, another there, and hands them to me. We get the idea (anathema to the student archaeologist in me) that we may help ourselves to a few. There are thousands and thousands of them. A rippled rim shard, a pot base with cord marking, a curved "shoulder," a little hollow tripod leg. Zhen keeps knocking them out of my hand with his trowel: these are not any good, later there will be better ones. Now a lesson. He takes a stick and draws in the dirt, in outline, the main pot types we will be seeing on these sites. He lays down the shards he has collected on his drawings and we photograph them. The video camera has not miraculously recovered. Our most interesting day will not be recorded on video tape.

There are two main pot types: the three-footed *li* and the round *fu*. The *li* was for cooking food (the three legs kept it upright and above the fire). It was incised before firing with a cord wound round a paddle. The small round *fu* was also cord marked. Li thinks that it may have been used as a measure for salt. Both pots are perhaps from about 1600 BC, and thus contemporary with the Shang Dynasty, a thousand years earlier than the site we saw in the Wu Valley, but much later than Neolithic sites discovered in Wushan County.

The Neolithic Era was not a package but a process and it did not occur everywhere at the same time. While Mesolithic and Paleolithic cultures are fairly uniform over immense areas, Neolithic cultures are not. These people were contemporary with bronze age cultures and may have traded a few pieces of bronze but they lived a Neolithic life.

Neolithic means literally "New Stone Age." Most people associate the onset of the period with the beginnings of food production based on domesticated plants and livestock breeding, ceramics and sedentary life. "New stone" refers to ground stone tools. These are generally symmetric, perfectly smooth, and shaped like axe-heads, about sixty cm long. They require great skill and an immense effort to make. Many have a hole bored through them to lash them to a handle. The other distinctive tool of the time, the sickle, is made of many small skillfully-shaped points mounted in a curved wooden haft.

Archaeologists, as I have said, talk in percentages, and especially changes in percentages of particular artifacts found in different strata. Crude stone choppers, like the ones we are finding here, were quickly made on the spot, with a minimum of skill, and did not need to be lugged from one hunting ground to another. They were still used at this quite late site, when they served the purpose, but they start to decline as a percentage of stone tools found. Stone arrow heads were used until they could be replaced with bronze in very much later times.

A few decades ago it was thought that pots were invented by farmers to store their grain. Archaeologists and ethnographers now agree that pots were first used to cook plant and vegetable food, grains and nuts, by pre-agricultural peoples. When the largest animals died out at the end of the last Ice Age (about 10,000 years ago) people became more dependent on plant foods. Seeds are much more palatable to old people and children, when boiled and it is easier to boil water in a pot than in a bark container with hot rocks.

Zhen is off now, through the fields, inspecting the terraces close to the river. He points out a small rise of land where the bank has fallen away and undisturbed strata can be seen. Li has also gone off by herself, using her trowel to pry out specimens, putting them in a plastic bag. Back she comes, to show Zhen and me a retouched stone chopping tool made from a river cobble.

In 1990, the authorities decided to investigate our next stop, Zhongba. Zhen had been with the team from the Sichuan Museum who did the salvage archaeology there. The site is 5000 m^2 but only 48 were excavated. They got C^{14} dates of about 2500 BC. There is no agreement yet as to a name for the culture: experts compare it to Hemudu, or describe the earliest levels as a Yangsho-like culture, or a variant of Daxi Culture, or suggest a new name, "Ganjinggou Culture." The tools were stone and bone; archaeologists cannot confirm that these people cultivated rice. Fishing and hunting were important; fish bones are especially prominent in their remains.

Even in ancient times, this area produced salt. A diet based on hunting and gathering has enough natural salt. Only in the Neolithic, when people became dependent upon cereal crops (millet, rice, wheat), did they have to eat salt to stay healthy. Sichuan was once at the bottom of an inland sea and so has rich deposits. Salt was mined by digging a shaft into a seam of salt or by drilling a well, dissolving the salt in water, lifting the brine out and evaporating it. "*Gan,*" from our first site and the name of the river, is a "mixture word": *Ganjing* is "brine." Before the recent reforms, local farmers still drew salt from the ancient wells. However the wells are now too deep, over twenty metres, and traditional mining is not commercially viable. Children have thrown stones into the old wells.

The people of Zhongba had domesticated water buffalo, pigs, dogs and sheep, as far as can be told from the bones that were excavated at the living site. Some bronze items, including weapons, were found but these are perhaps a bit later, from the Shang period. The pottery is similar to what we saw at Ganjinggou: sand and grit tempered, thick, fired at a relatively low temperature, cord marked, reddish. These, Kailin says, were not good conditions for making pottery; the kilns were little more than depressions with fire on top and bottom.

Zhongba is in a broad valley spanned by an immense bridge, indicating that the flood level is high. The water is very low now, exposing vast flats of sand and gravel and water-smoothed river cobbles. The archaeological site is on a large island in midstream, and it included a living area, a village. Zhen draws a diagram of the strata of the site, from Neolithic to Shang times.

Zhen then takes me in hand, leading me across the muddy flats: Robert falls considerably behind, photographing the bridge, the valley, and some workmen cutting stone. For Zhen, it is a little like coming home. He points out broken pieces of pots similar to those we had seen at the other site. Then he can't resist: he takes off at a terrific pace, his trained eye looking among the cobbles for ground stone tools that have been washed out of the bank by this year's floods. They form secondary deposits once again, but indicate what awaits future field seasons. Kailin and I clamber after him, doing our own shard hunting. He climbs high up the opposite bank, pointing with his trowel to exposed strata, packed with shards. He pries out some almost complete tripod (*li*) legs. I assume that this is a midden and that there will only be scraps. If there were any complete pots, the local people would have gone after them.

This pottery is coarse, tempered with grit, and fired at a low temperature, and, if I can tell from the shoulders and body shards (there are some complete bottoms), not very large. It is not beautiful: there can be no comparison with the thin painted pottery of Banpo or of Daxi cultures. No ornaments have been found and they mostly used stone choppers. And yet within a few minutes, we find two or three polished fragments, just by surface inspection. Zhongba, Zhen says, will also be flooded.

There is an old salt well. It is now five metres deep, dry, and lined at the top with cut stone. Originally the lining was wood. We return to the bus through a typical Sichuan village: a single street of old houses winding along the riverbank. Each house has a storefront, although only a few have taken off their wooden shutters because it is not market day. Robert is spellbound by ballet-like movements of two men making a cotton mattress on a wooden frame.

As usual we are behind schedule. I struggle to keep up with Zhen and Li, and Kailin limps along as best she can. I motion frantically to Robert and Zhongling to catch up. Stop and look by all means but "move it" between stops! We do not want to miss the famous Han tombs for a second time. We arrive back at the bus to face the mini-crisis. The driver has decided that he is very angry (his day, he says, is running late) and that another 100 is required (total 400). He demands payment in full, now. There is nothing to do but to agree and to explain to our Chinese friends, for whom this is a large sum, that it is worth it to take a little longer and to be able to take good pictures on such a beautiful day.

The next site is the Eastern Han tombs on Dragon Hill, near Tujing Creek. The site is important because few undisturbed tombs of the short-lived and little-known dynasty survive in Eastern Sichuan. These tombs are from the later part of the dynasty, roughly contemporary with the early Three Kingdoms Period. Kailin's translation is "cliff tombs" but our architecture book calls them "cave tombs," commenting that they predate Buddhism and Buddhist cave temples, and that they were "all the rage in Sichuan" in Han times. They have been found in the tens of thousands. Although many are much more elaborate than the Tujing tombs, the Tujing finds are significant. Zhen says that they help fill in some gaps in understanding relations between the Gorges and Central Plains areas.

Luo Erhu writes that these tombs were made by Han people and, although there are differences with Han tombs on the Central Plains, their appearance in SW China coincides with the influx of Han people in the Qin and Han dynasties (Luo 1988). Generally these tombs, like the stone pillars, reproduce the features of Han wooden architecture, an important reason for studying them. The entrance shafts were dug into the rock and gravel of the hillside and lined with cut stone blocks. A stone-covered drain ran their whole length. The tomb chambers, however, were hewn from solid rock.

There are 15 tombs here, of which we will see two. Tomb 5, a double-chamber tomb, was found intact in May 1981, creating a sensation in Japan and America, according to Zhen. (The excavation was written up in *Wenwu*, 1985(7).) The team leader was Zhang Chaijun from Sichuan Cultural Relics Management Bureau. The group included local members from the Wanxian and Zhongxian county relics bureaus.

We pile out of the bus. Before us, in the noon-day sun, is a hillside, a road, and a beautiful valley with Tujing Creek flowing through it. Zhen tells me that we will need our flashlights now. We cross the road and start climbing the slope beyond it, walking carefully so as not to harm the sweet potatoes. This time I am not so lucky or not so careful. I slip down into the muck, grazing my arm. No blood, so no matter.

We stoop through the grass and bushes, flashlights in hand. The opening to Tomb 6 is about a metre high. The entrance shaft and drain were removed by the excavators. Originally this door had been sealed with two stone slabs but they had long ago been taken away for their own use by the farmers. A short tunnel leads to the antechamber which is about 3 m wide and high enough for us to stand in the middle. The tomb proper was hewn from solid rock. The ceiling is decorated with a herring bone pattern, incisions in the rock about 6 cm apart similar to the cutmarks the peasant stone cutters leave today. There is otherwise no decorative carving, no painting, no inscriptions.

Beyond the anteroom are two burial chambers, one each for a husband and wife, large enough for a coffin and grave goods (This is called a three-chamber tomb). The local village had grown rich from duties on salt so this couple may have been very wealthy. The tomb was dated from two piles of thin bronze "cash": coins with a large square hole in the middle so they might be carried on a string.

The tomb we see is completely empty. Fu Kailin explains that in the Han Dynasty, people were buried in one of three kinds of coffins. Stone was most desirable, brick or wood were often used in the countryside. Wood was most common in the Three Gorges. Inscribed jades, placed on the stomach and the chest, usually told the story of the occupants of a tomb but no jade was found at Tujing.

Only Tomb 5, about five metres below Tomb 6, had not been robbed. Its entrance is smaller and full of dirt and garbage and we can neither go in nor see very far. Here, in a single burial chamber, they found the remains of a man and two women, buried in wooden coffins with 156 articles including bronze, silver, ceramics and porcelain. Pottery models had been placed near the door or in the anteroom. The excavators found a retinue fit for the gentry: servants, dancers, a flautist, a zither player, drummers, a fishermen, and a cook. The dead family had horses, dogs, pigs, chickens, carts, a well, and a cook stove. They are a priceless record of life in the area in Han times. We had seen some of these models at Shibaozhai, the only local artifacts we will see on our trip. The rest of the finds are in the Beijing National History Museum, and Chengdu's Sichuan Provincial Museum.

These tombs, Zhen says, will be flooded. They are at 190 m elevation, according to our report. Robert and I want to know where these people lived. Was there a town nearby? What was the relationship between the Han (conquerors) and the local people? We think this area is similar to northern Europe, and that the Han are like the Romans, living on the edges of Empire, ruling the fractious Britons. Our analogy evokes no enthusiasm in our Chinese friends. It is true that although the Han Dynasty fell, unlike the Romans, the Han people had been here eight hundred years and would not leave. The only thing that was clear was why people would choose this lovely valley for their tombs.

Li is impatient with our questions. There are articles on these sites and she will try to get them. But translation takes time and money, I explain. We are trying to learn as much as possible while we go, while we are experiencing things. Li's view, once again, is that this history is little understood and she is reluctant to give us information that might be inaccurate. We do not know enough about Chinese history to know that she is right and we stubbornly put some of this down to her Confucian habits of deference to authority. Zhen returns us to the bus, and asks the driver to take our picture with Li's camera. Our archaeological adventures are over for the day. We stop for lunch in the small agricultural town of Ruxi, where we will take our leave of the smiling and helpful Mr. Zhen.

We stop to see the wooden threshing machines and immense bamboo baskets in a farm implement store. The restaurant, to my surprise, sells bottled water. From here on we save our plastic water bottles and I make it one of my responsibilities to fill them from the hotel thermoses each night. As the boiling hot water will collapse the bottles, I choose a thermos, remove the cork to let it cool, hoping that the hotel staff will not notice. Bottled water is starting to be popular in China although our friends see little use for it. The thought of all those plastic bottles is as terrifying as the styrofoam boxes and the disposable chopsticks. The cost of providing cold drinks for everyone would be many Three Gorges dams.

We are thirsty, hungry and in high spirits. Our delays and the ¥400 are forgiven and forgotten. We are joined, as is customary, by everyone that we have picked up on the way. Beer and cola appear and we have bought their stock of bottled water. Cook and waiter are consulted, dishes are ordered: a magnificent fish Sichuan-style, a chicken including the feet, pork and green peppers, and soup.

Zhen tries to answer our questions about the tombs. He has a mustache and a ready smile, and a curious way of flinging his hands about and clearing his nasal passages while he talks. He agrees with me that western archaeologists might be very interested in the Mesolithic and Neolithic of the Middle Yangzi and especially in the development of agriculture. He hopes that his next excavation will be at a site called Minyan.

Does he think there are many more sites to be found? What about Early Man sites? Human teeth have been found at Wushan, at the famous dragon bone hill. There are many sites and, with the open door policy, the conditions for doing archaeological work are improving. While it is beyond all human power to do or pay for everything that might be done to understand the prehistory of the area, a large scale survey of the 600km reservoir is being conducted by the National Bureau of Relics, including an aerial survey.

Zhen is the first keeper of relics to say that help would be welcome (perhaps Li has suggested it, despite her caution that we are "just tourists"). They are ambitious, but lack money; they would like to cooperate with foreign experts. There is a document setting out a plan for excavating a thousand square metres, at a cost of ¥300,000 RMB. Could we get a copy of it?

Zhen thinks our book will be incomplete without some of the famous poems on the area. (How does he know about the book?) We say fond farewells. We have greatly benefited from Zhen Xianling's generosity with his time and his knowledge, and from his enthusiasm.

Back in Toronto, PhD-student Chen Shen and I locate an article summarizing the survey of Wanxian County by the Sichuan Institute of Archaeology in 1987 (*Kaogu* 1990(4):314-321). The team found 18 Neolithic sites: nine of them with enough similarities to be classified tentatively as a new local culture in Eastern Sichuan. The article also describes the work at Zhongba and Ganjinggou; only the latter has been extensively excavated. Someone has even suggested these sites may be early Ba-Shu. As a matter of course, the report calls for more excavations.

The day is still very fine and Robert and I do our best to take pictures.

A woman gets on the bus and the driver's helper collects money from her. A man gets on. This gives us a chance, through Kailin, to ask their opinion of the dam. As usual, they speak enthusiastically about the economic opportunity it will bring.

The crushed limestone road is mostly one lane wide and there are many switchbacks and hairpin turns. Large rocks have been placed on the edges, to slow traffic and warn of soft shoulders. We have one particularly close encounter with a truck piled high with white bags of salt, and I come eye-ball to eye-ball with a momentarily terrified woman, riding on top. We are also terrified: we are driving on the outside edge.

The landscape is very changeable. Some valleys are broad, others are narrow and deep. The hillsides are terraced from bottom to top. The lowest land is given over to rice paddies (but not at this time of year) and the much poorer higher levels grow sweet potatoes. It is clear here that bottom land is of much higher quality than land higher up and that peasants cannot be resettled simply by being told to move up. Peanuts are being harvested and spread to dry on flat round bamboo baskets like the ones we saw in Ruxi. There are beautiful bamboo groves and great dome-shaped haystacks, made of bundles of hay tied to trees. Sweet potatoes grow everywhere. Sweet potatoes roasted in a huge steel drum are the cheapest warm walking-around snack food you can buy in winter, and there is a dish made with boiled sweet potatoes and walnuts that is a great favorite in Sichuan.

Stone cutting is a major industry. Huge blocks have been neatly shaped and piled close to the road for pick up. There are many small brickworks. Across the valley workmen are digging a tunnel for a small hydroelectric dam (many advocate building many small dams instead of one big dam). Beyond is a pottery factory, its wares piled high into glossy pyramids. This is, still today, salt mining country. In fact, over mountain and down valley, and far off into the distance, snakes a brine pipeline, comprised of two elevated green plastic pipes. We do not know if the brine has industrial use, or if natural gas is used to boil off the water. There are no cement factories or coal mines and the countryside is clean and bright green, despite the time of year.

Many areas are wild and rocky, with striking geological formations. The tops of the mountains are crowned with mesas. Kailin has been asking the difference between a crag and a cliff. I answer that a crag is a particularly wild promontory: like the place Heathcliff went to meet Cathy's ghost. Then I realize a bit sheepishly that there is no reason to assume she's read *Wuthering Heights* or seen the Laurence Olivier movie. Her English is so good, I keep forgetting the differences in our respective stores of literary allusions.

Huge stone retaining walls have been built across the slopes to protecting the road from falling stones. As further testimony to their industriousness, as if any is needed, local farmers have dumped fill behind the retaining walls and planted sweet potatoes!

One of the towns we pass boasts much new construction. Apparently, it is the site of a new regional airport. The bus driver is now proudly our guide. This is the development area for the new city of Wanxian, the largest city between Chongqing and Yichang.

Isabella Bird, who walked through the mountains north of Wanxian all the way to Chengdu, noticed Kailin's crags. "The wealth of vegetation is wonderful. Not a barren or arid spot is to be seen from the water's edge to the mountain summits which are the limits of vision. The shiny orange foliage, the dark formal cypress, the loquat and pomegranate, the gold of the plumed bamboo, the deep green of sugarcane, the freshness of the advancing grain crops, and the drapery of clemantis and maidenhair on trees and rocks all delight the eyes. But the uniqueness of the neighbourhood of Wan consists in the number of its truncated sandstone hills, each bearing on its flat top a picturesque walled white village and fortification, to be a city of refuge in times of rebellion. These, rising out of a mass of greenery, with a look of inaccessibility about them, are a silent reference to unpleasant historic facts which distinguish Wan from other cities" (1899:175). We do not see these fortifications, perhaps not knowing to look, but I am sure that investigating them and their "unpleasant" history would be worthwhile.

Fengshui and Chinese Traditional Architecture

(Robert, On the road to Ruxi, September 19, 1992) As our minibus climbs over the spines of one hulking ridge after another my appreciation for the countryside grows. Each new valley reveals yet another complete set of terraced hillsides, precisely managed streams and carefully placed compounds containing farm buildings with black tiled roofs. Riding through the sun-drenched hills of Sichuan, I begin to have some understanding of the profound effects that the Three Gorges Project will have on the culture of the region. That limited understanding, however, is perplexing more than comforting.

When our rented bus sets out from Zhongzhou in the morning we are already carrying, in addition to the five of us, our guide Mr. Zhen, the driver, an older man who may be the driver's father and seems to be the boss and two other men who may also be relatives along for the ride. Along the way we pick up a few other people who were obviously known to the bus family. Some of the these passengers ride for a while and then get off. Others stay. I suggest jokingly that since we have rented the bus we should charge these folks a fare to help defray our expenses. Before the day is done, our hospitality will extend to more than the ride.

The best land is clearly furthest down the slopes. There the terrain flattens out in places to allow fairly large paddy fields. Some of these are dyked and used as fish ponds or have huge lotus leaves reaching up from the black water. Others, in this season, still have stooks of rice straw from the previous harvest. Occasionally a tethered water buffalo and calf graze on these fallow terraces. Buffalo are also being used to plough and prepare others for the next crop, probably wheat. I find it hard to imagine, given the small size of the fields, how any machine could improve on that method of tillage. Further up the hillsides there are garden plots bursting with green vegetables of various types, peppers, Chinese cabbage and what I would call chard. Interspersed with these patches are fruit trees, sometimes planted along the clay dykes that separate the paddies into their different levels. Further up still are sweet potatoes, peanuts and no doubt other things that I can't identify.

Although seldom seen from a passing bus I know from our closer encounters that each farmstead has pens for pigs. Chickens roam the farmyards and sometimes venture onto the road. In one village I see dozens of ducks sitting on the side of the road, tied together in groups of three by one leg each. They are waiting to be transported to a nearby town. Later I see a cluster of them hanging out of a bus window, feathers flying in the wind.

At one point the road hugs the inside curve of a small valley set in one of the hills. We stop beside a stream that is dammed by a curved stone retaining wall. There is a graceful little stone-arch bridge over the stream. On the smooth rocks under the bridge several women are doing laundry. Outside each of the houses clustered around this crossroad there are wooden racks on which noodles have been hung to dry. Everywhere I look there are noodles, miles of them. In another village on the road to Wanxian there are expanses of exposed flat bed rock. The area may be 10m square. Spread on the rock to dry is recently threshed rice.

The remarkable fact about this agricultural mix is that it seems to be able to supply the entire range of produce for the regional diet. Little, if anything, is imported from any distance at all. Furthermore, all of the activities and sequencing of crop and livestock production and of food processing (noodle making for example) are interconnected like the proverbial Chinese puzzle.

Caroline and I want to stop and talk to farmers but that isn't going to be feasible. Even though we have rented the bus it is only for the work day and time is flying. Also, while we are clearly foreigners, the fact is that our educated city friends are also somewhat out of their element in the countryside. Kailin has pointed out a number of things of interest during some of our walks through other paddies. But her's and Zhongling's periods of banishment during the Cultural Revolution are not regarded with the kind of nostalgia that some North Americans and Europeans have for their summer jobs as farm hands. They have a very limited desire to stroll into the paddies and translate conversation about crop rotation. So I am left with my brief observations. Still, some patterns are clearly discernible.

The puzzle fits together roughly in this way. Wheat for making flour for noodles is planted in some of the same paddies where the staple rice is grown. The rice occupies the space from planting in late spring to harvest in fall, while wheat is planted in autumn and harvested in spring. One kind of vegetable follows another on the higher terraces. But between the successive crops, which in much of Sichuan can be grown for twelve months of the year, some of the paddies are used to pasture animals. The droppings from the ducks become an important element in the growth of eels which are released into the paddies while they are flooded. The fish then mature while the rice grows in a practice which some say had its origins in the Three Kingdoms period almost two thousand years ago. The trees growing along the margins of the paddies also drop material which nourishes the fish and plants, while sediment from the paddies can be dug out to feed the trees. The roots of the trees in turn help to consolidate the clay dykes that separate the paddies on different levels. Animal as well as human manure is another vital part of fertilizing the hard-worked soil.

The land, of course, does not work itself. The people who live in these valleys have not only moulded the land to suit their needs but have created living and working places as well. But the rural vernacular architecture is in no way grand, self-conscious or even permanent. It is said that the loss of a house, due to accident, natural disaster or simple disintegration, is of less concern than the loss of the food that might be stored inside. There are relatively few towns in the Sichuan mountains and the villages are quite small and spread out. The settlements rarely have many public buildings other than schools and some offices. Those buildings don't differ much from the other rural structures. Traditionally even temples are rare since most rituals are carried out in the home.

The meaning of the buildings, therefore, is derived totally from their utility. The subdued aesthetic that they do possess is rooted entirely in everyday life. The buildings are often built in a manner that will allow them to be added to. The builder begins with a rectangular shape to which first one wing and then another can be added to form an L and then a U shape. The buildings are simple, usually symmetrical, and in spite of the repetition of form and material I didn't find the older-style ones at all monotonous.

From their very foundations the older rural buildings reveal themselves as being a kind of transitory reshaping of the elements in the immediate environment. In every way they come from that environment and when they have served their purpose they return to it. Older buildings generally have floors of compressed earth. There may be footings of stone at the corners or a stone wall, laid without mortar, surrounding the base. But this feature is more to stabilise the earth floor or to serve as containments for poured concrete in the newer buildings than to act as a base to fasten down the walls. Those walls, of which there are basically two types, are, in fact, not secured to the foundation at all. One style of wall is load-bearing. These are made of tamped earth (compacted soil), mud brick or fired bricks. In this design the roof structure is supported by the walls. The alternate type of construction consists of pillars, usually of wood, that support the roof. When the second style is used, the areas between the pillars are filled in with bricks, stones, wooden lattice or some combination of these and other material.

Of the different types of buildings the ones that are probably most surprising to Westerners are those of tamped earth. They are made by placing a mould consisting of two parallel boards on the base of the building. The mould is two or three metres long, with the planks perhaps 30cm deep and set about 30cm apart. Soil is dumped into the mould and packed down. The mould is then moved up and the process repeated in successive layers until the desired height of wall is achieved. The process is thousands of years old.

Virtually all of the building styles that we see in the Yangzi Valley are transplants from the north Chinese homeland of the Han people, who long ago conquered and settled the area. In more northern conditions earth walls are practical, but here in the south, adaptations were required. During the heavy rains that fall seasonally in Sichuan it is possible for a tamped earth building to literally melt away. Although all types of walls in rural Sichuan have been protected by plaster and whitewash when circumstances permitted, the most universal response to the challenge of rainwater has been the development of broad overhanging eaves.

The earth and stone for walls and the clay for brick and roof tiles have traditionally come from the soil virtually underfoot in the mountains of eastern Sichuan. We see many places where clay is being excavated from patches right among the paddy fields. Kilns are built along the sides of roads and fired by locally mined coal. Sometimes roof tiles are made in the form of cylinders about the size of a saucepan and then skilfully broken in four pieces to yield the arc-shaped tiles that are interlocked on the roofs. That method is said to date back to the Western Zhou Dynasty, perhaps one thousand BC, but the only tiles I see being made have been moulded into their final shape before firing. Neither the bricks nor the tiles are of high quality, but they seem to serve their purpose quite adequately.

The largest number of buildings we see are probably fairly recent in origin, two storeys in height and of grey brick construction. Nevertheless the more traditional styles are very much in evidence. While we do see tamped earth buildings both close to the Yangzi and in the valleys leading back from the big river, timber-frame construction is more common among the traditional-type buildings. These have an appearance not unlike their half-timbered counterparts in England and Germany, which is what gives so many of the Sichuanese villages their mediaeval feeling in Western eyes. In the past the timber used in construction was reasonably plentiful. For reasons that become increasingly apparent to me, the availability of wood is being seriously diminished. Judging from the look of many of the timbers

I suspect that, as in other parts of the world, serviceable building materials have been used over and over in successive generations of structures. The same may be true of stone and bricks.

It seems that the timbers used in the construction of the roof supports in these vernacular and in the monumental buildings in the Yangzi region, are much larger than those used in the West. They are also put together in a way that requires each structural member to carry more weight over its horizontal length than is the case with the more delicate Western style of roof truss. Yet another difference can be seen in the fact that the timbers used in the Chinese method of construction are seldom squared. I learn that all of these aspects have reasons. The Chinese believe that the heavier a building's structural members, the more stable the building will be. Therefore, roof timbers are left in the round to add weight, though not necessarily strength, and the structure is made as massive as possible. Rafters, also left in the round, are placed over the purlins that join the roof frames. To complete the heavy roofs, the rafters are covered with planking and that is overlaid with a sort of plaster. All this serves as support for arc-shaped clay tiles that are laid in alternating and interlocking rows to shed the rain. We see only a few thatched roofs. The building methods are dictated by tradition and governed by mystical beliefs. But given the facts that the whole structure simply sits on its base and that China has experienced many earthquakes over its history, they are quite reasonable solutions to the challenge of building.

The persistence of timber as a construction material in China is somewhat puzzling to Westerners since so much of our building tradition is invested in stone and brick. First of all we have to realize that it is easy to fall into the ethnocentric trap of assuming that our way is normative. It is similar to asking why the Chinese use a logographic rather than a phonetic style of writing. The question isn't which one is right or even better but do they work in their milieu. It is not that the Chinese never mastered masonry. They were building magnificent stone bridges and multi-storeyed brick pagodas a thousand years ago. Timber, however, has been available, is relatively strong, durable when maintained, easy to work with, to mass produce and standardize. Furthermore most buildings, especially rural ones, have always been relatively small. Land for agriculture is too valuable to cover with buildings. The general absence of a priestly class and the lack of a religious tradition that demanded the glorification of other-worldly deities have made the development of large buildings less an imperative than in Europe and the Middle East, In fact the Chinese philosophy of harmony with nature has been well served by modest wooden buildings. Timber worked very well for them, and the Chinese simply chose not to build in other fashions.

The ordinary rural buildings in the middle Yangzi region can be seen to fit well into their environment both in their use of locally available materials and in their style of construction. Their placement in the landscape might also be viewed as demonstrating a remarkable degree of environmental sensitivity. For governing purposes rural areas are divided into "administrative villages." But these are purely constructs, areas defined on a map. What one sees are "natural villages" that seldom conform to the government divisions. There are at least two clearly discernible types of these villages in the valleys through which we travel on our way to Wanxian. One is a centralized collection of buildings in the bend of a river or at a crossroads. The other is a series of buildings strung out in a line along a ridge, road or stream bank. I'm sure these places average less than a hundred residents each. Scattered between the villages there are also a few isolated farmsteads which probably house family units. Collectivization, in force in this region up until 1984, does not seem to have altered the traditional settlement pattern very much except that some of the buildings were used as work unit offices, clinics and other centralized facilities.

At first, to a Canadian used to settlements built on grid patterns, these Chinese villages seem to be a chaotic jumble without a plan. Thinking of European towns turns out not to be of much help in making sense of China either. While older European settlements are also seldom arranged on grids, their locations can often be understood in terms of their defensibility, such as being on hilltops. I see few villages in China that looked defensible to my, albeit rusty, military eye. The larger towns are invariably in places of strategic importance and always sited to take tactical advantage of the surrounding terrain — but not the existing rural villages. Still I am sure that the organization of space is meaningful to the people who live there and I was anxious to try to understand that meaning.

China is not the oldest civilization in the world. Metal refining and other supposed measures of advanced culture were probably going on in the Middle East well before they were in the Middle Kingdom. I have already begun to realize as well that there are few if any old structures in use, by North American, let alone European standards. What China may lack in antiquity, though, it certainly makes up for in continuity. There is a sense, which I feel at the time and later will confirm in my reading, that settlement patterns and building types in the areas we visited have been relatively unchanged for centuries. This fact is both exemplified by and, to some extent, explained by the practice known as *fengshui*.

Literally translated *fengshui* means "wind and water." Various English or techno-English words have been used to express *fengshui* including: geomancy, divining form from the earth; topomancy, divining form from the landscape; astroecology, studying the relationship between the living earth and the stars; and mystical-ecology, relating the living systems of earth to human beliefs. More straightforward terms such as topographical siting and just plain siting have also been tried. Because the discussion is about a deeply rooted complex of ideas, however, most writers prefer to simply struggle with understanding the term itself in the original.

I am beginning to realize that the Chinese are not religious in anything like the way people in the Judaic, Christian or Islamic traditions understand the term. But they certainly are spiritual and, especially here in
the countryside, that understanding of the universe and humanity's place
in it has been consistently expressed in the careful placing of structures.
That expression has gone on, one building or one tomb at a time, over centuries and perhaps millenia. Traditional Chinese culture seems to have the
kind of holistic or ecosystem view that we in the West are just rediscovering.
Celestial or other-worldly forces and terrestrial or earth-bound elements,
aspects of *yin* and *yang* in Chinese terms, are understood as having to be in
balance. The placement of buildings, therefore, must acknowledge the need
to maintain that equilibrium. In a peasant society where the land is all-im-
portant they talk not of building but of settling the earth. Over hundreds of
years of formal writing and accumulated folk tradition, a systematic canon
as well as a profession has developed which guide the practical decisions
that impart specific pattern to both rural and urban Chinese landscape.

As always it is a dubious and potentially misleading exercise to attempt to capture an intricate and highly developed practice in a short explanation. But some of the elements involved are: orientation to the cardinal
points, the configuration of hills or prominent points and the relationship
to water. Generally people prefer their houses and other buildings to face
south. But in the complicated micro-climates of the Sichuan and Hubei
mountains there seem to have been many variations on this theme (Zhang
Fei Temple, for example, faces north). With regard to surrounding terrain
features, or *sha*, it is considered best to have high land behind and lower
eminences on the flanks and in front. Each of these prominent points has
a name and is ascribed particular characteristics. The symbol of a dragon,
or *long*, is used to impart a kind of mystical and, at the same time, anthropomorphic dimension to the landscape. Its body is seen to run through the
setting in a certain orientation. It is said that, when water flows in wealth
flows in, and so the placement of villages on streams is very important. This
is more than just the convenience of not having to carry water a long distance. One of the consequences of *fengshui* is that the patterns of settlement in widely different places have a familiarity and cultural resonance.

In some ways it is possible to see *fengshui* in quite practical terms.
Placement with reference to cardinal points and prevailing winds usually
provides the village with good cross-ventilation, protection from the worst
weather and optimum capture of light and heat. Location on hillsides keeps
the buildings off prime agricultural land. Care with regard to drainage
usually means protection from flooding and adequate water supply for
farming, drinking and fire-fighting. In these ways *fengshui* represents the
accumulation of environmental wisdom. But I think there is a danger in
reducing our understanding to a functional common denominator. What is
important in this area of rural China, and I suppose elsewhere as well, is

not specific buildings, as we would think of in the West, and not even definable regional styles, but the pattern of life itself on the landscape. Material culture here is not a collection of things so much as the way things are done. To appreciate the *fengshui* of rural Sichuan, we need to adopt the level of abstract thinking that we might bring to the viewing of an expressionist painting. The picture is not of a person but of what the person is doing.

On yet another level, land use patterns probably have, embedded in them, long-term ownership configurations, elements of the former feudal organization and the attempts of the communists to reformulate land tenure since Liberation. All these aspects are overlaid and intertwined in the ongoing rhythm of life on the land. The present villages and townships are descendants of the communes which in their turn replaced the feudal holdings that came before. The transitions from one system to another, however, have not ever been easy. Modern planning techniques have already clashed with *fengshui* in the Chinese experience. There are examples, some of which we see, where urban and external patterns have been introduced. Some of these were efforts to impose a more rational order on settlement. Some were imposed for ideological reasons such as collectivization. In 1984, however, the central government gave up its efforts to force collective agriculture on the rural population. This has given way to what is called "household responsibility." Some of the collective farm buildings we see are abandoned, their painted Maoist slogans fading away like the remnants of other, older dynasties. My suspicion is that they had been built without recourse to that accumulated local knowledge which *fengshui* embodies and that they were simply in the wrong places. They were probably too cold in winter, too hot in summer. There is now reported to be grudging recognition by planners for the efficacy of local and traditional beliefs.

My attention is suddenly caught by a construction project on the far side of the valley. It is not clear to me at first what I am seeing but as various pieces become visible at different turns in the road the complete picture emerges, It is a small hydroelectric generating station. My interest in hydraulic and hydroelectric power installations, began with my work on the history of the canals in the Niagara region of Ontario, the site of some of North America's first power projects. The fall of water in this case might be 30m or 40m, and the stream is of modest size but appears to be of consistent flow. On completion the project might yield ten or 20 megawatts.

First we see the powerhouse. At other intervals up the valley are portions of the partially completed penstock that will carry the water from a reservoir to the turbines. These in turn will run the generators. Near the top of the valley there is a dam about five metres high. All three elements, dam, penstock and power house, are being constructed of local stone. The style of construction is quite traditional. The sedimentary rock breaks up conveniently into manageable pieces when blasted. Each piece can be manhandled by one person. The stone is being laid with mortar. The front of the dam consists of a series of curved bays with substantial buttressing between each bay. The penstock seems to be flat on the bottom, straight-walled and arched over top. It is large enough for a person to walk in upright and it snakes down the side of the valley, often set into a ledge that has been excavated to receive it. Parallel to the penstock in places, and running a couple of metres above, there is evidence of an old, and perhaps even ancient, road. I can see at one place a wall with a kind of doorway in it.

"There," I point. "That wall, what is it?" Zhongling strains to see it and then points it out to our local expert, Mr. Zhen.

"I'm not entirely sure," says Mr. Zhen in translation, "but I think this was a place where in olden times a rich person made travellers pay to pass by." A toll-gate, I think, but I am surprised that for all his interest in below-ground, prehistoric archaeology, Mr. Zhen isn't clearer about historic ruins such as this one. The more I see in that part of China, however, the more I realize that ruins are quite rare and therefore not of nearly as much interest in historical interpretation as they are in the West. In China, any place that is inhabited and used is built on over and over again. For that reason, and owing to the general use of evanescent building materials, ruins of older structures almost never survive. Only in an inaccessible place, such as this road remnant on a ledge, would a ruin remain. But if exacting road tolls was the way to make a living here in the past, the valley's value today is in the fall of water.

Some commentators on the Three Gorges project estimate that power equivalent to the promised output of the one big dam on the Yangzi could be derived from harnessing all the potential on the smaller streams such as this one. I don't know if that is true or, in fact, can both schemes be accomplished together? What can be said about a project like this one is that the environmental impact, even of many such plants, would be considerably less drastic than the one big dam at Sandouping. A pond of the size being created in this valley will be more easily accommodated in the local agricultural and settlement pattern. The construction materials and methods are indigenous. The big dam, by contrast, will require massive resources and expertise from outside the region and even outside the country. It is not that the Chinese are new to significant water-control projects. The *Dujiangyan*, or All Rivers Weir, in Guanxian, Sichuan, is a remarkable water-diversion installation which was first built in the third century BC. Like the terracing of the hills we are driving through, it has taken shape gradually from generation to generation. The large dam and water-control projects undertaken since liberation, on the other hand, have sometimes taken place suddenly, without time for the ambient culture and economy to adjust. The Three Gorges project will be the most disruptive yet.

At the top of the highest valley, before we begin our descent towards Wanxian and the Yangzi again, we stop for lunch in a market village named Ruxi. Market villages are one of the rural planners' attempts to encourage people to stay in the countryside and not drift into the larger cities, as so many millions have done in other developing countries. These enlarged villages provide commercial and industrial services to the surrounding countryside. The main street, or rather the only street, in Ruxi sports some fairly new two-storey brick buildings which have been plastered and white-washed. One of these houses a shop of farm implements, including brand new, hand-operated grain winnowers of the type only seen recently in Canada in agricultural museums, but which I remember as a boy in the 1950s being still stored on my relatives' farms.

Beside the farm implement place is a restaurant with wicked-looking peppers spread in heaps on the sidewalk out front. I know I am about to eat some of these deadly veggies drying just centimetres away from the ubiquitous open drain and that, sooner or later, food prepared in this way is going to get me. The lunch, however, is as wonderful as usual.

Nothing is actually said by way of invitation, but when the bus stops everyone piles out and goes in. Our group now also includes the bus driver, his father, his brothers or cousins and the other passengers we happen to have with us. I don't count the number but we fill two large tables, which is all the restaurant holds. Lots of beer appears immediately and the mood is festive. Mr. Zhen, who is by far the most entertaining person we will meet, continues regaling us with jokes and historical anecdotes that lose little in translation. He says he could speak to us in our own language but he made a terrible mistake as a student and studied Russian instead of English. Plate after plate of food appears. Fried vegetables, pork, and a geriatric, bony chicken with a sauce that redeems it from death by old age. Finally the fish arrives triumphantly, its head and tail reaching for the sky.

These remote valleys produce all of this. Our guests and passengers are hard-working people with little by way of hope for material rewards. But our lunch is a cause for genuine celebration and good feeling. Eating, which I have already concluded is the central cultural expression in Sichuan and the activity toward which all the delicate balance of building and farming in these valleys is directed, is a pretty fulfilling activity. The bill for this extravaganza comes to ¥64 (US$12.80). Before Mr. Zhen leaves us, we give him some of the cigarettes I've been lugging around. He laughs and says that he will not be bribed by foreigners.

Luck: A Fat Pig With Yuanbao (Money) on His Back

(Kailin, Fuling, September 19, 1992) Like all travellers, we had a habit of looking at doors, windows, stoves, pillows, everything. In Shibaozhai I found lucky patterns of great vitality and interest. I felt, owing to China's reform and openness to the world, that more and more people are in high spirits, full of hope.

Chinese believe in the power of images. They believe that if they are surrounded by objects decorated with images embodying wishes dear to the heart, it is more likely that such wishes will come true. The local people used auspicious symbols to meld meanings, feelings, and humour into auspicious patterns, expressing wishes for longevity, happiness, good health and prosperity.

In Chinese art, there are four basic methods of communicating ideas. The first way is where a message is expressed directly on an object. For example *"shou"* (the character for longevity) might be glazed on a bowl, written on a fan, carved on chopsticks.

The second is a more subtle manner where the messages are communicated by motifs derived from flora and fauna, mythology, legends, historical events and geometric designs. Some symbolize attributes that are closely associated with the intrinsic characteristics of the subjects that they portray. There were a lot of vendors' stands at the ports along the Three Gorges. I found the local people like to use lucky patterns of animals. Sichuan bamboo chopstick heads were carved with flowers and birds. The most interesting was a magpie standing on a plum tree, predicting that good fortune would come. These chopsticks were also used for hot pots in many restaurants. Because bamboo is a fast-growing plant, reaching maturity in less than a year, it is cheaper than wood, and harder, so bamboo chopsticks are more easily disinfected. In a shop in the small town near Zhongzhou, I found a lovely and humorous papercut design of a group of mice posed as bride and groom, with the bride in a sedan chair, in all her finery. The mouse was the ancient mark of a family having grain to spare. I saw a child's tiger-head hat, tiger-head shoes, and a lotus hat when I was travelling with my mother last August. Tigers are fierce and protect babies from the five evils.

In the third use of images a motif and an inscription can be combined. The inscription is added to clarify or enhance the symbolic meaning of the motif. At Fuling, I saw a birthday cake in a sweet shop on which there was, in cream, the character *"shou"* on a background of a giant stone from the Yangzi River. The giant stones also symbolize longevity (that you would live longer than the stones of the Three Gorges). Motifs acquire new meanings by a play on words. The Chinese language is monosyllabic. As a result, punning is widely used in symbolic art. For instance, when the character *fu* (happiness) is posted upside on a door it means "come," signifying the advent of happiness. The sound *"lu"* means both deer and a handsome salary, so sometimes the picture of a deer is used as a good luck wish. The lotus flower and fish imply "have something to spare every year." "Lotus" (*nian*) harmonizes "year" (*nian*). "Fish" (*yu*) harmonizes "to spare" (*yu*). The Chinese "rose" (*yueji: yue* means month, *ji* means season) and "ancient bottle" imply "safety all the year round." The pine tree, or the bamboo in the cloud, or the narcissus imply a "bumper harvest."

The fourth use of images features a variety of door gods and decorations that convey the message that life can be rich and colorful, and brightened by means of art that is close to home. There were so many door gods that I didn't recognize them all. Some said *"Yucijingde"* and *"Qin Supao"*; others said *"Zhang Fei"* and *"Guan Yu."* The most attractive lucky pattern was the "New Year's Day" painting on the door: a fat pig with a lot of *yuanbao* on its back implied the family has a lot of grain and pork to eat. It predicted a good harvest for the farm house in the following year. *Yuanbao* was a shoe-shaped gold or silver ingot used as money in feudal China. On Qiling Street in Yunyang, we saw pictures on doorways of a lovely child riding a *qilin*, a unicorn. In some paintings, a sacred book hangs from the *qilin*'s horn; in others the *qilin* is accompanied by the goddess and her maid servants, who send children. One antithetical couplet on a doorway read: "The child of *qilin* is in heaven, the first scholar is in the human world."

In Chinese legend, *qilin* is a fictitious animal that is tame and kind and symbolizes benevolence. It has a history of at least 3,000 years. The legendary *qilin* is said to have the body of a deer, the tail of a cow, the neck of a wolf and the hooves of a horse. Its fur is yellow. It has a horn on its head for defense, and walks rhythmically without treading on insects or damaging the grass. It makes its appearance when the country is prosperous.

In China's earliest collection of folk songs, the *Book of Songs*, there is a poem in praise of the *qilin* that likens its kindness, cleverness and benevolence to that of a ruler in the 11th century BC. According to another legend, when Confucius was born, a *qilin* appeared in his hometown and spit out a sacred book. The story implied that although Confucius was not on a throne, he had moral integrity and was to be addressed as "the Most Holy Sage." Of course this is purely a fabrication, and it did foster in later generations the feudal, male-chauvinist concept of connecting *qilin* with the birth of a son of the family.

Once the great poet Du Fu (712-770 AD) praised the two sons of a Mr. Xu, one for being handsome and the other for being energetic. He said that the boys were sent by Confucius and Sakyamuni, and were angels of a *qilin* in heaven. At Shibaozhai when a Chinese peddler saw the *"Lao Wai"* (Caroline, the foreigner), he asked ¥100 for a small pendant *qilin* made of aluminum. This was only because of the foreigner's high-bridged nose. I bought it for ¥70. The peddler explained at great length that the *qilin* was an amulet which dispelled evil. Caroline said the price in Beijing was ¥150! Perhaps because of legends it became more valuable.

Also in Shibaozhai I saw for the first time the *taotie*, a ferocious mythical animal with an enormous mouth. *Taotie* was an imitation of ancient bronzes. I remember my father laughing at me: "You are eating like a *taotie*." The word is used to describe a greedy person as well as a voracious eater.

The Chinese are also fond of the dragon and the phoenix. The Confucians considered the dragon, phoenix, turtle and unicorn the four auspicious animals bringing luck to the emperors. Along the Three Gorges, they are found everywhere, on primitive pottery, on bronzes, ships, vehicles, clothing (printed on T-shirts), hangings, ceremonial articles, utensils, musical instruments, furniture, ceramics and coins. In Confucius' temple at Fengdu there was a colourful painting entitled, "The dragon and the phoenix embody good luck." On an inscription at Baidicheng, a pair of phoenixes were dancing. On old palaces and temples, dragon figures coiled around pillars, slithered down roof eaves and flaunted their coils on magnificent screens. The dragons and phoenixes on the roof tiles at Shibaozhai gave the illusion of flying swiftly skyward. No wonder Du Fu once wrote that, "Those who associate with dragons and phoenixes will have irresistible power." Later, however, the phrase "associate with dragons and phoenixes" became synonymous with "playing up to people with power and influence."

In Baidi Town there was an interesting historical note about Liu Bang (256-195BC). It went something like this: Liu Bang, who founded the Western Han Dynasty in 206 BC, was said to be the "seed of a dragon" since his mother had borne him after dreaming of copulating with one. The lion (*shi*), a symbol of power, often stands guard (*shi*) in front of temples. The two characters make a pun that is understood by the reader. So there are always two lions, side by side, meaning "everything goes well." Lions are often depicted rolling a ball made of strips of silk, indicating that a good thing lies behind them.

In looking for folk art, one looks not for spectacular expressions, but for the regular and the commonplace, confirming the ordinary, rather than celebrating the novel. A culture is kept vital by the harmony of the members of a group, not by the efforts of a lone individual. In the Three Gorges, communities and families keep folk arts and crafts alive. Thus, a wide variety of unique local patterns and designs express best wishes for good luck, happiness, kindness, longevity, good harvest, love of loving beauty, joy and most of all, hope and confidence in the future.

Sustainable Development

(*Robert, On the road to Wanxian, September 19, 1992*) It is hot and the sun is still bright as we start the long winding trek down to Wanxian. We look soberly at the beautiful terraced fields, fish ponds and bamboo groves around the brown villages. Most people are resting in the swelter of the afternoon but a few can be seen labouring in the paddies. There is always a danger, of course, in becoming sentimental about the simple lives of people in picturesque places. Foreigners who make short visits away from their own situations of comfortable consumerism can be guilty of wanting the people they see to be the character actors in bucolic fantasy. But we should try to see these situations not as our picture, but in their own true historical and broader cultural context.

Yet cultures are never simple. There are at least four strong traditions in Chinese history that are competing here in the shadow of the Three Gorges. There is the age-old urge to mould the land for human needs. There is the deep and almost mystical agricultural cycle. There is the dark, ominous cycle of catastrophy that has been the pulse of Chinese history. And there is the lure and promise of the modern city and the good life.

The Chinese landscape that we are viewing, after all, has been totally manipulated by human activity and moulded over centuries to serve human needs. There is very little left here that can be called natural. Is the idea of building a huge dam across the world's third largest river only different in scale and not kind from the terracing of these hillsides into a myriad of rice paddies? The difference, I suppose, has to do with evolutionary or incremental change as opposed to revolutionary or sudden change. The paddy fields in these valleys have been built over dozens of generations. The flooding of the reservoir will happen almost instantly by comparison.

About a million and a quarter people will be displaced. It is reckoned by some to be the largest planned movement of people in history. We are assured that everything will, in fact, be well planned. But I wonder what the new villages will look like? I wonder, in fact, where they will be since we haven't yet seen any land that isn't already occupied. One proposal suggests that the people be moved to border regions far away from Sichuan. Another report I've read says that the farmers can move farther up the slopes of the valleys and farm there. But we can see clearly that every available square metre of space that can be won from the mountainsides and put into cultivation is already being exploited. The best and most productive land that has sustained life here for centuries will be lake bed.

Many of these ideas for relocation come from the Yangzi River Planning Office. But that agency is located hundreds of kilometres away from here in Wuhan. I strongly suspect at this point that the planners from Wuhan have never been here. It seems like a strong and radical charge to make, but our own city friends don't like being in the countryside very much, and I doubt they would ever have come here if it hadn't been to accompany Caroline and me. One might think, however, that it would be a vital part of the planners' job to come and see the places they are talking about. Then I remember a story told by a woman I met in Hong Kong. She was conducting courses for a United Nations agency in China, teaching professionals about community work. They were supposed to go out and undertake an actual project at the end of the course. They were paid to do it and given film to record their experiences. But when they returned, most of them had simply treated the time as a vacation. The pictures were mostly of the trainees visiting tourist sites. Few of these people had ever been out of the cities before and they didn't really want to deal with the rural folks they'd been sent to help. Perhaps my suspicion is not altogether unfounded.

What strikes me about these valleys is that the flooding will not mean simply moving people, shifting production and seeing some villages disappear. It will mean the complete disruption of the culture of the area since that culture is expressed primarily through the complex interaction of seasonal labour devoted to food production on land that has been managed, up until now, in a delicate balance with nature.

There is much talk in the world these days about the concept of sustainability. The aspiration is to find a way to meet the needs of today without compromising the ability of future generations to provide for themselves. It seems to me on this passage through the long-settled valleys parallel to the Yangzi, that the life-style of these people may be about as close to sustainable as one can find anywhere on earth. The climate is benign, water plentiful and the soil rich. There is a system of agriculture which provides all the food necessary for the regional diet and there is a built form that seems to fit into the environment in a balanced way. By continuing to carefully husband all the resources at hand, life here could conceivably go on indefinitely.

There are a couple of catches. The first is that, by modern world standards, the life-style of the people is very modest. I wonder if this modest, non-consumer way of life might in fact be the only one that is sustainable? But another factor imposes itself on my thinking in the form of a fleeting and almost insignificant vignette. Just as I am waxing rhapsodic in my thoughts about the balance with nature and the potential, in theory at least, of agriculture and simple life being sustainable in rural China, I happen to see a man walking through one of his paddies with a tank strapped to his back. In his hand he holds a hose running from the back-pack tank. He is spraying what I assume is pesticide. As well, we have seen plenty of evidence of chemical fertilizer. The point this brings home to me is that while agriculture might potentially be self-sustaining in these valleys, the fact is that it is being pushed to the limit, as agriculture is everywhere in China. The tremendous need for production has led to the introduction of such things as chemical growing agents and chemical pest control. This means that cash is needed to buy these items and industrial facilities are needed to produce them. The land, therefore, is no longer being worked in perfect balance with its capacity. There may be little choice in a country with such an enormous need for food, but the carefully balanced production cycle is broken and ultimately will not likely be sustainable without increasing inputs of artificial agents.

All that of course is happening quite outside any consideration of the Three Gorges Dam. These conditions of stress on the agricultural countryside are general. So too is the situation of excess population. The combination of a maturing baby boom, the increased efficiency in production brought about by the dissolution of communes and a return to more entrepreneurial farming have led to a growing redundancy in the rural population. It is estimated that as many as thirty per cent of the peasantry are no longer needed to maintain agricultural productivity. The drift of these people into urban areas is inevitable.

I also needed to remind myself that Chinese history has been punctuated with recurrent disruptions to the lives of these rural masses. Each dynastic succession every few hundred years has been accompanied by chaotic periods of war and famine. Intermittent periods have seen peasant revolts, such as the Taiping uprising in the last century. The population of Sichuan itself was decimated by turmoil at the end of the Ming Dynasty, about three hundred years ago. Most of the people here now are the descendants of immigrants from other parts of China who repopulated the region.

The disruption that will accompany the flooding of these valleys will not see massacres by roving armies or starvation, as in the past. Its most likely outcome will be increasing urbanization. But this urbanization of the rural poor is not necessarily considered a bad thing. In fact there is a strong association of city life with wealth. For the last couple of thousand years the vast majority of Chinese people have lived on the land, as they still do. The minority who have been urban dwellers have generally enjoyed a higher standard of living. Forgetting that in the past the privilege of urban residency was strictly controlled and that the wealth was derived primarily from rural productivity, many people welcome the chance to move off the land.

Nevertheless the reshaping of nature in China is an integral part of this culture. Perhaps the present period of social disruption brought about by the flooding of these valleys adjacent to the Yangzi is gentle and humane by comparison with the catastrophes of previous centuries. Maybe the benefits promised are worth the price since the area, however beautiful to us, is a tiny portion of China? Providing gainful employment and living places for these and other people who will likely migrate into the cities is a huge challenge for Chinese planners.

These are not questions that I can really answer. I am fairly sure, however, that whatever the rewards of the great power and flood-control project at the Three Gorges, very little positive impact will be felt by the people in the valleys between Zhongzhou and Wanxian. There might be other farm land for them somewhere, and there might possibly be some urban jobs. Even so, it is unlikely that these particular people will ever be among the principal beneficiaries of future developments.

The bus is now descending fast. The villages are becoming larger and more frequent, the valley broader and more populated. As we come around a bend the bus slows to pass a construction site. A hillside is being excavated to widen the road. Zhongling and some of the other men on the bus fall into an animated discussion. Through the clouds of dust I see dozens of workers. All have shaved heads, are shirtless and are wearing black trousers rolled to the knees.

"Are they prisoners?" I ask Zhongling. Caroline turns around and passes me a hand signal that suggests I drop the subject. This is probably one of the reasons why travel by foreigners has, until recently, been so restricted. Officials here like to be sure visitors see only what they're supposed to. Penal labour has become a contentious international issue. The old communist idea of re-education through labour had its admirers and apologists in the West. Sending selfish landowners and corrupt government officials out into the rice paddies to learn humility and to gain an understanding of the lot of peasants was an approach that might be defended. But now there have been accusations of Chinese prison officials using virtual slave labour to manufacture goods for overseas trade. If we were alone with our friends they might have been able to be more open but, in company, perhaps Caroline is right to warn me off a discussion. It is Zhongling, however, who continues.

"What makes you think they're prisoners?" he asks me.

"They're all dressed the same and their hair is cut," I venture. "And I don't think even poor peasants could be persuaded to do this kind of work in the heat of the day."

"You're right," he says, but that is the end of it. He continues to talk to the other men in Chinese. They seem to have definite ideas about the scene and also some embarrassment at our having witnessed it. I remember my mother guiding some foreign visitors around Kingston, Ontario, where we lived when I was a boy. She was similarly reticent at their questions about the prisons that are so much a feature of that city. So I am left to wonder to myself whether those men were political dissidents or common criminals. Were they students from Tienanmen Square or crooks who'd pilfered funds from the collective?

5

WANXIAN

Taibai Delights

(Caroline, Wanxian, September 19, 1992) We enter Wanxian about "half way up." Compared to other towns along the river the traffic patterns are modern. The streets are broader, tree-lined, and some have formal sidewalks. A few people ride bicycles. Storefronts are neatly tiled and there is a look of hustle and prosperity. Robert theorizes that it was bombed in the War and since rebuilt. The driver drops us at the by now legendary (in our minds, at least) Taibai Hotel. No one seems to speak English, there is nowhere to change money, but it does have the look of a one star hotel! The men are assigned a double on the "men's" floor, the third, and we women a triple on the fifth. The water is hot, the toilets work, there are towels and complimentary tooth-brushes. The toilet seats are firmly in the right place. Taking no chances, this time I put myself first in line for the bathtub.

My roommates, as a matter of course, set about hand washing their clothes and hanging them all about. As well, they hang out the clothes washed yesterday, but put away wet. Not so below. The ever resourceful Zhongling ascertains that this wonderful hostelry has a laundry service and once-worn jeans, shirts, even wool jackets, are put (at expedition expense) into the obliging arms of the charming laundresses. Robert does half and half: half he washes, half he hands over.

There is yet more washing to be done. Li unpacks the potshards from her plastic bag, scrubs the mud off them with her complimentary toothbrush, sets them on a clean towel and then places them on the table in the bedroom to dry. By now I have begun to worry that we will be unceremoniously asked to leave, and mop up the worst of the muck and the puddles on the vanity counter and the bathroom floor. Hastily I explain that the lamp table's finish could be ruined by water. *China Daily* makes another appearance and the potshards are spread on it on the floor.

It is now 5:00p.m. I hatch the idea that my video troubles must be due to a nearly-spent lithium battery. Perhaps we can find a replacement, now we are in a big city? But first, Zhongling must finish in the bath. Hurry up! Zhongling and washing, Zhongling and laundry, are becoming major causes of comment. With Kailin's foot still sore, Robert and I need him to come with us.

We hail a taxi, the first we have seen since leaving Chongqing, only four days ago. To a video store. ¥10. Anything. And so we careen down the mountain. No luck. No camcorders are sold in this town. Zhongling worked for many years in a precision instrument factory. He prods the prongs that hold the battery down. A helpful young woman in the store suggests we test it. It tests fine. Zhongling asks further directions, and we set out on foot for a department store. There goes the easy answer to the camcorder problem. Robert promises to look over the manual tonight.

What's next? The big plastic bag has given out and a backpack like mine or Li's must be found to replace it. Kailin, despite her sore foot, has been carrying the bag much of the time, so the new bag is to be hers. So, off we go, from one luggage store to the next. Although there are many goods, there is little selection and quality is hard to find. We settle for a blue denim bag marked "made in Hong Kong" — but we doubt it.

We now have only ten minutes to get back for dinner at six. We pass by one mammoth set of stairs only to be directed to another. This one twists and turns so that it seems that there is just this one last flight . . . . It is lined with free market stalls. People are doing their shopping on the way home from work. Tomorrow is their day off. By the time we reach the hotel, I am almost prostrate.

There are some pointed comments (from me) about the double standard on the laundry. (Mostly I am angry because they didn't call up and tell us about the laundry service!) Well, the guys say, why not put your things in the dryer and have the staff iron them? Well, the laundresses have gone home for the weekend as Sunday is their day off. Later it becomes clear that there is no dryer: the men's freshly-laundered clothes have been hung to dry on the roof. The roof is locked and that is that. So our laundry continues to drip about our room. Next, we learn that the staff member who fetches boat tickets has also gone for the day.

We had been asked, on checking in, if we were going to eat at the hotel. Reluctantly, we had said yes. It is somewhat past the appointed time when we arrive in the dining room to find a table set with a dozen cold appetizers. We hasten to point out, through Kailin, that we had not agreed to a tourist-style feast. I am not fond enough of these smoked pig parts, pickles, and eels to actually pay for them. Finally they are removed and we set about finding what is still available. Dinner is not a success: the fish, when it finally comes, is all bones; Three Delicacies from the Ocean includes a lot of Spam. Invariably, the "best" places seem to have the worst dining rooms, although we did arrive late, often a fatal error in China.

Except for Zhongling, who sets out to buy some oranges, we have all had enough. Our room is damp (very damp) but comfortable. Kailin offers to do my washing. I refuse: in Canada we do our own washing (or we send it out). She insists that her offer is out of friendship and she likes to keep busy. I suggest, instead, that she wash my potshards. This wrecks the third complimentary toothbrush and contributes even more to the humidity. "They are not as beautiful as Li's," she says, mercilessly. And "What am I going to do with them?"

Getting them has been so unexpected that I really have no idea, except that perhaps Norman's famous family or my immense skills at handling things in China or, more to the point, my great good luck, will solve the problem. There is nothing rare about them and if I had not picked them up, they might have been ground into nothing by a farmer's foot. I realize that I am temporizing. In the meantime, they are added to the pile of plastic water bottles, snacks, *China Daily's* and other things to be put in the new blue denim bag.

I take out John Keightley's *Origins of Chinese Civilization*, brought all this way, and show Li some of the illustrations of Neolithic pots from the lower Yangzi, drawn up and organized into neat typological sequences. Some of them are similar to the shards we have found today. She shows me that she is reading a Chinese translation of Canadian anthropologist Bruce Trigger's *Time and Tradition*, an important book on the development of complex societies. I explain that he is an expert on the Hurons, the native people of Canada who had built the village where I had been a student excavator this past May. Only then does she realize that the pictures I had sent her of the dig meant that I had actually done some archaeological work. Kailin's translations and Li's very basic English do not go so far as discussions of Trigger's book, which I had not in any case yet read, but Li takes a new interest in me. I am interested that Trigger has been translated into Chinese.

Kailin is not to be left out. This afternoon she had given the bus driver a tape of popular film songs to play on his tape deck, and had sung along, with a much better than average voice. It is not any good now, she says, but nine years ago she gave a concert. Once more I am made to realize how much individuals lost out in the struggles over the years. But, Norman would say, they also gained a lot. Even Kailin.

Where are our oranges? Where is Zhongling? I had hoped we could call Norman before setting out on the river again. Zhongling finally shows up, calls the desk, they put through the call. Norman is no better, no worse, but he is going out to the university to see some old friends tomorrow.

Town Planning

(Robert, On the road to Wanxian, September 19, 1992) We are now coming into the outskirts of Wanxian. It is the largest centre that we will visit on our trip between Chongqing and Yichang. It is also a place whose past brings up sharply some of those issues about Westerners and our history of interaction with the Chinese. The so-called Wanxian Incident was the local version of a scenario played out in various places in China beginning with the Opium War in 1839. In a series of lop-sided armed conflicts in the 19th century, China was forced to open its markets to Western trade. Since there was often little from the outside world that the Chinese wanted, Britain and other nations encouraged the use of opium in China so that they could then supply the drug and take valuable Chinese goods such as silk and tea in return. The Chinese retaliated by growing their own. This middle Yangzi area was one of the last places that was forced to accept foreign trade and in the 1890's opium was the biggest crop. One of the main products of the valleys in these Sichuan mountains, was tung oil, derived from the nut of a local tree. The oil is a very useful wood- finishing product, prized in the West until it was largely replaced by synthetics.

It was in Wanxian in 1926 that a dispute arose between the local authority and the foreign powers who were intent on controlling trade on the river. The events and even the way we talk about them are indicative of what has been wrong with Western attitudes toward China. In foreign accounts of the Wanxian Incident the Chinese antagonist is referred to as a warlord, a term with a pejorative connotation in the West, and indicates complete disdain for people without civilian-controlled central governments. It also has racial overtones. In the 1990s an African country such as Somalia has warlords, while the former Yugoslavia has presidents, generals and spokesmen. In our own Western history we don't refer to Cromwell, Bonaparte or Robert E. Lee as warlords. Most European and New World countries have gone through periods of disunity and decentralized control, as well as military rule and civil war. It may have happened more recently in China, but in the long run of history, it may well have happened less often. I think it is significant how we choose to talk about history.

At any rate, while there were American, French and other gunboats on the Yangzi in the 1920s it was the British who were principally involved here. Wanxian had only recently been opened to foreigners. The Chinese were accused of interfering with a foreign steamship company trading in the port. Two Royal Navy vessels steamed up the river. Something went wrong. The British began bombarding the city with their big guns. The shelling started fires. The British also machine-gunned anyone they could see on shore, assuming them to be the troops of the local army. When it was over there were three thousand people dead in Wanxian. The foreign newspapers in Shanghai reported that Western honour had been upheld.

The Chinese have good reason for being a lot more sceptical than they are about the value of things foreign. As it is, I think the genuine friendliness which we encounter on our travels is quite remarkable. But while the China of the 19th and early 20th centuries often received outside influences reluctantly and even under duress, the China of the 1990s seems to be anxiously, even recklessly embracing foreign ideas. It is against this background that I observe Wanxian. It is not only the largest community that we visit, but the largest that will be totally relocated.

In recent Chinese economic and regional planning there has been an attempt to foster the growth not only of rural market towns, such as the place where we had lunch, but of small and medium- sized district centres such as Wanxian. This provides an alternative to the uncontrolled expansion of a few major urban areas, which is a condition that prevails in many or even most other developing countries. To some extent the Chinese have been successful in this effort. Unfortunately, however, the system may be breaking down. It is estimated by some observers that anywhere from ten per cent to thirty per cent of present urban-dwellers may be illegals. That is, people who have not been officially assigned to work in the particular city in which they actually live. Recently there have been reports of many country people, who arrive illegally in Beijing and other large cities, being sent back to the provinces. How well small cities such as Wanxian, and larger ones like Chongqing and Yichang, cope with the influx of people who will be displaced by the Three Gorges reservoir is a critical factor in the future of the region. Part of that success hinges on how well the communities plan for expansion.

Town planning may or may not have been invented by the Chinese but there is little doubt that they have been masters at it for a couple of thousand years. It begins with the same strong and well-established principles of *fengshui* that characterize village lay-out and the orientation of individual rural houses. The *fengshui* of Wanxian is said to be almost perfect.

> ... for a distant range affords protection from the evil influences of the *Yin* or darkness; a lower range on the opposite bank of the river screens it from the south, yet is low enough to allow the benign influence of the *Yang* or light to penetrate. Finally, a smooth point of rock juts into the harbour just above the city, thus providing the indispensable Lung or Dragon. ... (Worchester 1971:493)

To *fengshui*, however, the Chinese can add an almost equally long tradition of specifically urban planning principles. Han Dynasty cities of the third century BC were being structured according to certain formulae that have recurred over and over and given shape to most of urban China. For the most part Chinese cities are rectangular in form, with regular grid patterns of streets always meeting at right angles. Four main ideas dominate. Cities are generally a series of walled enclosures; a central axis predominates; the axis is invariably oriented north and south and the individual structures within the city are formed around courtyards. But as with most things in Chinese culture these principles originated in the northern plains. They are suitable for the topography there and to many areas where they were carried by conquest and migration.

The cities that grew in these mountains however, couldn't possibly have followed the regular grid pattern. They don't spread over a flat countryside like Xian or Suzhou: they cling to precipices. They don't occupy land so much as they defy gravity. As a consequence the streets are not parallel but follow natural contours and naturally occurring ledges. In all of the towns we see there are roadways that have either been wide enough for vehicles in their original form or have been widened to accommodate modern trucks. Coexisting with the roads is a network of stairs and ramps that carry only pedestrian traffic. In fact a large portion of the goods transported up and down is still carried on human backs. I see people carrying everything from baskets of fresh food to a refrigerator slung from a pole between two men's shoulders.

Any idea of axiality or north-south orientation is rendered impossible by the exigencies of the local terrain. Most of the towns once had walls, but these are almost all gone now and can never have had the concentric form of the rectangular cities. Few houses in these towns have room for courtyards, although some manage simulations of them. Here, urban builders, having attempted to accommodate the mystical principles of classical Chinese planning, have been forced to adapt their forms to suit the landscape.

Wanxian is a good case-study for what might happen in the future resettlement of the people in the Three Gorges area. Whatever damage was done by the British bombardment in 1926 was repeated by the Japanese during the Second World War. Consequently certain parts of the city appear to have been more recently planned and rebuilt than most of the other places we visited. There are many examples of the old-style roads, interconnected with complexes of stairways, some grand and wide and some little more than alleys with steps. At the same time there are some broad boulevards with treed medians, a park featuring a very Western-looking clock tower, and a large traffic circle, with fountains and a statue in the centre and streets radiating out. What Wanxian represents, then, is a place in transition between the adaptation of traditional Chinese urban modes to the mountain and river terrain on the one hand, and the modern influences that are associated with a dependence on motor vehicle transportation on the other. To these matters of street layout and traffic patterns are added another contrast to traditional Chinese urban form. The multi-storey building is a very recent introduction with no history in Chinese vernacular architecture.

I wondered what the new towns that the government proposes to build will look like? There are, of course, many aspects to urban planning, but just some of what I see gives me cause for serious worry. The model suburbs of Chongqing's new economic zones signal the wholesale importation of Western urban planning. There are wide highways with limited access. Arterial roads connect to the highways and minor streets snake off the arterial roads in the typical North American crescent and cul-de-sac pattern. The high-rise buildings are set back from surrounding roads, with the intervening spaces "landscaped" with trees, shrubs and grass. When I ask some planners about services they say there are underground sewers. I see what I take to be storm water retention ponds. Finally there is evidence of the principle of land use separation whereby workplaces, commercial zones and residential complexes are all in different places.

But what is wrong with all that? Why shouldn't the Chinese be "modern?" "What is suitable urban form for a given area?" and "What will work in the local society?" From this point of view I see a number of things that I don't have any hesitation in labelling as unfortunate.

To begin with, the North American approach to road networks is extremely wasteful of land. We have even begun to resist it here, where land has been relatively plentiful. In China, in one generation, almost one-third of all arable land has been lost due to various forms of development. The land use pattern in North America has been described as very friendly to cars but not to people. There are ever-increasing numbers of motor vehicles in China, but for the foreseeable future of the Gorges walking will continue to predominate as the only way about for most ordinary people. The so-called new economic zones are nightmares for a porter transporting items on his back or on a carrying pole. In addition, the increased distances entailed necessitate more public transportation which neither governments nor individuals can afford. There are almost no diesel engines in China, the gasoline used is still leaded and there are no enforced emission standards. The environment can afford more vehicle traffic least of all. While planners may favour wide roads and dual highways, the local people are considerably more practical. In many places along the new highway between Chongqing and its airport, the peasants have simply commandeered the outside lane of pavement for drying grain.

The planting of trees and grass around buildings is primarily a European landscape-architecture style that emulates the country villas of the Isle de France and Berkshire. It is a further waste of land, outside the Chinese tradition of enclosed courtyards in towns and walled gardens connected to palaces. Furthermore, the rural Sichuan reality sees edible crops planted right up to the walls of buildings and to the roadsides and children grubbing in the dirt. The so-called landscaped areas around modern buildings will be difficult to maintain both because of the expense and the foreignness of open parkland to Chinese cultural traditions.

Western-style underground sanitary sewers in the Chongqing industrial suburbs, so I find out on further questioning, simply empty into the river, without either primary or secondary treatment. This presents a considerable pollution problem now, but it will grow to catastrophic proportions when the fast-flowing river is transformed into a lake. By contrast, in traditional, and even present Chinese towns, human waste is not a problem but an agricultural resource that is still more or less systematically collected and used to fertilize the land. Storm water in the modern city also drains into these new sewers. What appear to be storm retention ponds turn out to be Western-style landscape design features into which water must be pumped.

Land use separation presents another urban design minefield. In the past, Chinese urban communities were divided into a series of compounds. Traditionally these were occupationally organized around their respective guilds. There was a weavers' quarter, a metalworkers' neighbourhood, a potters' district and so on. People lived and worked largely within their own area. That ancient system was continued and even strengthened in more modern times with the establishment of factories and the imposition of the work unit structure. The work unit, consisting of dozens, hundreds or even thousands of people, took over a function similar to the ancient *fang* or walled city block.

The negative aspect of this social and work organization is, first of all, that people are forced to live in close proximity to industrial processes, such as foundries and chemical plants, which are often very noxious. These situations are analogous to the industrial revolution conditions in Europe that led to the concept of separating work and residential uses. A further problem with the Chinese work unit system is that people have been assigned to jobs at an early age and are expected to continue there throughout their careers. Workers not only live, toil and recreate within their work unit walls but are also provided with health care and other benefits without venturing outside. Retirement is provided for as well. Often children are expected to take over their parents' jobs. While this system provides many amenities and a degree of security not common in Chinese history, it also results in a sort of industrial serfdom.

There are, therefore, some very good reasons for the Chinese to look for new ways to organize their urban space. There is a genuine desire for more social mobility. People don't necessarily want to marry someone who has grown up in the same work unit, or work at the same job all their lives. Most of the friends of Caroline and Norman that I meet, already live in situations where one spouse works in one place while the other travels an hour or more to another work site. They have to do this because housing is still provided only through the work unit. From a management point of view, there is a growing realization that a free labour market can be a considerable advantage in increasing the effectiveness of work. For these reasons, separating work and living spaces is clearly desirable.

At the same time the new reality is that fewer industries are as polluting and as environmentally unfriendly as they have been in the past. In China, as elsewhere in the world, completely benign service industries are growing. So while there are some tendencies that might point to the benefits of breaking up the traditional industrial compound, the adoption of the Western separation of land use patterns is not the only alternative. There is an opportunity in Chinese urban planning to avoid the suburban sprawl phenomenon that plagues so much of North America. Perhaps it is not necessary to have everyone so far from their work that they need to drive or be provided with public transportation. Perhaps commercial and recreational amenities can be dispersed throughout cities so that shopping and entertainment don't also require long-distance travel. Perhaps the freeing of people from industrial compounds doesn't necessitate their enslavement to expensive transportation systems. A further factor is simply the magnitude of the demand for housing. Members of a 1970s baby boom are now starting families. The typical four to six-storey apartment blocks that we've been seeing will probably be replaced with buildings that are much taller and much more numerous.

If the design of the new site for Wanxian, and the other communities to be moved along the Yangzi, doesn't respond to the realities of tradition, terrain, social customs and economic practices, then the results could be disastrous. Moving 1.2 million people will be difficult and traumatic enough. But if they move into urban spaces that are as unsuitable as the new suburbs I see in Chongqing it will be a compound tragedy. I'm optimistic that the potential changes in many people's lives in the region might be advantageous. But I don't think those advantages will be realized if people are moved from established urban settings, where a pattern of life and work has evolved over hundreds of years, and are placed in spaces designed for another landscape, in another culture and for an alien economy.

Uncontrolled Urbanization

(Kailin, Wanxian, September 19,1992) Caroline and Robert seem to be very concerned about what they call "uncontrolled urbanization" during resettlement. This is because their consciousness and cultural background are different from ours. In their country, rich people have villas in the countryside, and the farmers are rich. Of course, they are not willing to leave the countryside.

But China is an agricultural country. Some farmers cannot solve the problem of dressing warmly and eating their fill, although a lot live far better than before. In their eyes urban people are in a better position. They try their best to jump out of the countryside by marriage, by replacing one of their parents in the city or entering university. Usually the children of farmers study harder and get higher marks in the entrance examinations for university. Joining the army is another way out. They try for promotion to the rank of cadre and by that means to enter a military university. People try many things to get themselves registered in a city or town.

Of course, with the Three Gorges Project, "uncontrolled urbanization" might occur at the very beginning. But if the government arranges migration systematically and solves job problems properly, the situation will get better and better. Robert has asked me what I think about the disruption of patterns of life. I say this is not a bad thing, but a symbol of China's progress. There is an old Chinese saying, "Men in motion will survive and be invigorated, trees in motion will die." Movement through space is related to economic opportunity. A far larger migratory stream moving toward cities implies the stimulus of superior economic opportunities. Perhaps migration will bring economic success, and promote confidence that society rewards men on the basis of merit.

Mao Zedong said, "The poor think more of changes." Usually poor people move about more rapidly. If we can understand the nature of the urbanization process it will dispel some common misconceptions concerning working class "ghettos." Because farmers can get money for resettlement – a higher amount than the value of their property – they will change their way of life. It is that when farmers lose their land, urbanization will change their patterns of life. But we hope the lot of farmers will improve.

Silk Fans and Pebble Art

(Kailin, Wanxian, September 19, 1992) The Three Gorges enjoys the reputation of being a world-famous art gallery. For artists, it contains inexhaustible source material for creation.

Folk arts and crafts today enjoy high visibility and homemade items are often presented as emblematic of the region. As the Three Gorges abound in bamboo, the "bamboo" industry is rapidly being developed. Basket making provides a good example. Although the various genres are found all over the country, the baskets of the Three Gorges have some distinctive features. Modern machines cannot produce satisfactory copies of traditional baskets. The most attractive basket, I thought, was one with a small bottom and big mouth. As well as baskets, craftsmen have specialized in creating bamboo flutes and toy whistles. Artists without conventional studio credentials have created highly innovative works. I liked the exquisite bamboo strips painted with a picture of the poet Qu Yuan best. Caroline liked a sewing box with a nice green pattern so much that she bought it. The mats made of thin bamboo strips are practical for our hot summers. I estimate the inexpensive mats here would have a good market in the cities known as the "three furnaces": Chongqing, Wuhan and Nanjing.

The regional arts and crafts tradition also includes chair and furniture making, weaving and spinning, beadwork, jewelry, rug making and leather work. I never thought Caroline would be interested in grass-woven shoes, but she bought two pairs of them.

Later in the trip when we visited Shuanglong Town on the Small Three Gorges, I saw a lot of imitations of ancient relics, bronzes and coins. Their workmanship was good, and the ancient people's work worthy of imitation. So contemporary artisans turn their talents to designs which may be a century old or older. They repeat rather than create anew so that traditions will survive. Their sense of maintaining the tradition is more pronounced than their sense of self expression.

During the Three Kingdoms, when Zhu Geliang plotted his military strategy, he carried a feather fan, expressive of his dignified bearing. During the Tang Dynasty, emperors presented their favoured officials with feather fans. In the Song Dynasty, feather fans were replaced by round or pleated silk fans. Fans have also been important stage properties. Male characters in Chinese Opera might use a big fan to depict a general, while a young lady might carry a silk fan to indicate her grace and beauty. Scholars opened or closed a fan to indicate their cleverness, while slave girls might use a round silk fan to show their charm and amiability. The legendary monk Ji Gong always carried a broken fan. He was a helper of the poor and his broken fan was to teach bad rich men a good lesson. Ji Gong always traveled, so people called him *Yun You*, which in Chinese means roaming or wandering.

Famous Song Dynasty poet Su Dongpo, in his work *Chi Bi Huai Gu* (Meditating on the Past at Chi Bi), described Zhou Yu, the commander-in-chief of the troops of Wu Kingdom as follows:

> The recently married Zhou Yu, dressed in a silk turban and weaving a feather fan, talked cheerfully and humorously, and all the while his troops were turning Cao Cao's army into ashes.

The earliest fan we know of from archaeological excavations comes from Chu Tomb No. 1 in Jianglin, Hubei, from about 2,300 years ago. Its shape is similar to another unearthed at the later Ma Wangdui Tomb from the Han Dynasty. The back was decorated with black and red geometric patterns of woven bamboo. The Sichuan-style fan prevailed in the Ming Dynasty. To increase its magnificence, craftsmen would concentrate on the handles and ribs. They would choose a valuable material and carve it delicately or inlay it with a mosaic of pearls and jewels. The most valuable fans were those where a famous calligrapher or painter agreed to write a poem or draw a picture on the covering. Emperor Zhao Jie of the Song Dynasty, a talented painter, wrote a poem in the cursive style on a round silk fan. Such famous calligraphers and painters as Zhen Banqiao, Ren Benian, Wu Chansuo, and Qi Baishi left expressions of their artistry on fans that have become immortal treasures.

The Three Gorges abound with paper and bamboo. Today fans are found everywhere. Frequently, fans are gifts given to relatives, friends and acquaintances as tokens of respect, affection, faith, or general goodwill. Although electric fans, refrigerators and air conditioners can surpass the fan in effect, people still carry a fan and enjoy its gentle breeze. Old men like to carry big palm-leaf fans, emblematic of their leisure and their freedom.

Fans are a local specialty of Wanxian. When we entered Wanxian, we plunged into a sea of fans of all varieties: sandalwood, ivory, paper, feathers. Local people tried to find an original approach. If the fans of Suzhou and Hangzhou are well known for their elegance, the fans of Wanxian shall be known for their size! The longest, largest fan I have ever seen is a 2.2m giant I found in Wanxian. The smallest is only 3cm long. The most attractive is one covered with the Chinese character for "longevity" called the *shousan*. When I asked the price, the salesman answered, "The Japanese offered a price of ¥10,000. It is already sold." A beautiful fan, made by a skilled craftsman, can be the visual and aesthetic centerpiece of a room. Pleated fans are often hung on the wall as a picture or used as the headboard of a bed.

A paper fan, about a third of a metre long, without calligraphy or painting, costs about ¥1 in Wanxian. The idea is to have it painted by or for a person you like. I presented Caroline and Robert each with one, inscribed, "United as one, we can stand against the strong gale." These were symbols of friendship. With its long tradition of usefulness and beauty, there is no reason to think that the fan will not fare well in today's market economy!

Plants grow slowly in the rough terrain of the Three Gorges and often their shapes are eccentric. Roots are found in a wide variety of colours including gray, white, yellow, red and black. Black roots are miracles wrought by water in nature, by the scouring of the restless Yangzi. These roots are the raw materials for the root carver's art.

Stones from the Three Gorges are also prized. "What did you see on your trip to the Three Gorges?" a friend asked my mother when we returned. "Besides stones, there was grass everywhere," my mother said. She thought that the mountains, crags, and shoals were stones, all stones. People like to associate the shapes of stones with something. Nature's weathering of stones is a kind of creation. The famous French sculptor Rodin said, "Beauty is everywhere. If we don't see it is not due to a lack of beauty but to our lack of finding it." The creation of beauty depends on inspiration; the discovery of beauty requires intuitive knowledge.

Pebbles, cobbles and shingle are a common sight all along the Gorges. Most tourists think of them as material for building houses or paving roads; few think of pebbles as works of art fit to grace the most noble palace. Artists have no need to carve: they need only put together, by shape and colour, their choice of stones. Some works consist of five or six stones, others two or three. For refined or popular tastes, the pebble artist uses his rich imagination and aesthetic sensitivities and ingeniousness. Usually the craftsmen will clean the pebbles and perhaps draw a picture on them with a brush and lacquer them with a transparent varnish. A work of art is born. I like the panda's necklace best and I bought two. In another a monk carries two pails of water, two monks carry a pail each, three monks have no water. They are not bad but they are too heavy to carry with me!

Yichang, which we will only visit briefly, is close to the primeval forests of *Shennongjia* Mountain. The movement of the earth's crust and nature's weathering have twisted the stones into grotesque shapes, creating the "Potted Landscape of Yichang." In the Villa of the Gathering Clouds more than three hundred potted landscapes of hills, rivers and tree stumps are on display. What I liked best was the replica in stone of Three Travelers Cave, called "the pearl at the mouth of the Gorges," and the miniature of Wanxian City, built up with tree branches and rockery.

All of these arts — of wood, bamboo, grass, roots and stones — have a strong potential for development.

Sunday in Wanxian

(Caroline, Wanxian, September 20, 1992) The comforts of the Taibai Hotel are much appreciated. Li as always rises early, silently slips out and finds herself breakfast. The sleepy heads find that breakfast at the hotel is long over. Alas neither public nor private enterprise offers Sunday brunch in Wanxian. It is hot and we are still tired and soon give up looking. What could be better than big white buns from a small bakery, mine with sweet red bean paste inside, instant coffee and early fall apples?

Robert and Zhongling volunteer to try to get boat tickets. I am glad to spend the morning resting and catching up on my journal. Robert has solved the mystery of the camcorder: simply, it has come to the end of the film. I cannot believe it. The weather is overcast and hot and the slightest activity makes me sweat. The air conditioner has been permanently turned off so we open wide the windows. The men return successful. Their good first impressions are confirmed: too bad this town will have to move. But the feeling in Wanxian is of excited anticipation.

It is already past noon. The desk clerk scurries to assemble our account: the dining room has forgotten to forward our supper bill. A couple of yuan are charged for damage (allegedly) done to the room below by laundry Robert hung out the window. Robert is indignant, but we pay. Thank goodness I saved the table! Then, just as all that is settled, the girls enter in procession bearing Zhongling's laundry, fresh and neatly folded, including the wool jacket. Our washing, still wet, is stowed in plastic in our bags, not so neatly folded.

We have no energy to go anywhere. The nearby Bank of China opens as promised, but not today to cash traveler's cheques. We must make do with the cash we have. And so we sit in the relative coolness of the lobby until it is time to go to the ferry dock. I had thought, after our success yesterday, that we might hire another bus to Yunyang, our next stop. However, our driver had warned that this road is much worse than the one we have just been over. I put little store in this warning, but the road winds far back into the countryside, it is hot, and yesterday was quite expensive. There is no amiable Mr. Zhen to show us archaeological sites. The boat is cheaper and quicker.

There is something Kailin must tell Robert. People are laughing at his ponytail and asking, is this foreigner a boy? The Promoter had seen him combing it and laughed. Why? We explain that in Canada men with ponytails are unusual. Robert says it is to indicate that he has an artistic soul, making it clear that he feels apart from "ordinary life." Zhongling is still baffled. The notion of the rebel artist is by no means unknown in China. The pigtail is also a symbol of Manchu oppression, and a source of ridicule from foreigners.

To our (illogical) delight, we are again on boat 110 (we have only been on the water three times!). By the time we get on board, my fatigue has turned into a headache. Robert and Zhongling have only been able to get fourth-class tickets. Fourth class is, not surprisingly, less comfortable than third. There are 18 berths in each cabin; no desk, thermos or basin. The linen is less elaborate and not so clean. But Zhongling has done well: my berth is next to the door. I can lie down and look out at the water, enjoying the bit of breeze and the motion of the boat, writing in my journal. Strauss waltzes blare from the PA system. Not bad for a few yuan.

The Wu Mountains are not as high or wild as warned but, in the relative coolness of this lower deck, I am content that we have chosen the river. On the stern deck, Robert and I are approached by a man with a good camera and a big lens, a bank employee from Chengdu. He has come to see the Gorges while they are "still unchanged." He speaks English quite well, and like most intellectuals, is enthusiastic about the economic reforms but nostalgic about the Gorges. We talk photography and Zhongling hovers in the background. Seeing the Gorges before it is too late, comes up again and again.

Never have changes come more quickly than in the last forty years. The government has dynamited rapids and shoals and installed navigation systems. Steel, steam and diesel have taken over almost completely from wood, winches, oars and trackers. The dangers that took so many lives and held back development are only a distant memory.

6

YUNYANG

Little Emperors

(Caroline, Yunyang County Town, September 20, 1992) It is late afternoon when we arrive in Yunyang County Town. The boat drops us off at the foot of the most daunting set of stairs yet. This time, we concede, a porter is required. Somehow, from the milling crowd, a very small man is chosen. Zhongling makes a pile of all of our bags, except for our precious cameras. I object. One man is to carry everything? Zhongling is twice his size. "Don't worry. He's used to it." Our friends do not share our North American squeamishness. The porter makes two piles of our bags, ties them with his ropes and fastens one bundle on each side of his bamboo pole. We are off.

He *is* used to it. I can hardly make it up the stairs. Our problem is keeping up with him and our bags. He is only minimally shod in cheap canvas slippers, but practice and calves and thighs of steel, make it possible to carry our load and keep his footing on the greasy stairs. Li asks directions to the best hotel. We turn left at the top of the stairs, walk, climb to yet another level, and turn left yet again onto the straightaway.

The Yunnian Hotel and Restaurant has only its pictures of bright, simple rooms to announce its presence. The women at the door are helpful. I gratefully sit down on the tiny lobby's one couch to await the outcome of the negotiations. The porter puts down our bundles and joins the confusion. Often it seems to us that the porter has just arrived in town and is just as interested in the sights as we are. A crowd gathers in the street. Kailin is dispatched to inspect the rooms. They are clean but there are only "Chinese" bathrooms down the hall. This newly-opened hotel is privately owned. Yes we are amiably told, there is the usual "hotel for foreigners" further on, and it does have ensuite bathrooms. I am falsely encouraged that greater comfort lies nearby.

The porter agrees to continue, provided we pay him more, and off we go. It is a few Chinese minutes away, ten more minutes, in fact. The entrance and stairs to the Lanxian Hotel are immense, but they are dark and filthy. We pass a crude and cluttered dining room. The lobby is empty except for one couch, on which I deposit myself, while Zhongling, as usual, attends to the negotiations and the paper work. It does not look good but it is getting dark and we are too tired to retrace our steps. We tip the porter and he leaves happy.

This hotel has also just been privatized. It is run by what seem to us to be a constantly expanding number of family members, capable of various levels of helpfulness, misinformation and menace. There is no wait for rooms to be made ready. We are immediately led, baggage in hand, to the fifth floor. Our rooms open off an outside corridor that is also a balcony. There is a breeze and a wonderful view of the teeming streets and the river. Just below is the roof-top patio-cum-Buddhist shrine of, we are told, a local capitalist.

As for the much-anticipated bathroom, the lights do not work, the water is off, and the toilet seat must be retrieved from under the wardrobe. The beds are numbered 1, 2, 3, meaning that this room is usually rented, if it has ever been rented, traditional Chinese style, bed by bed. There are mosquito nets and that seems ominous. At the far end, however, is a barred-in balcony with a lock — and a clothesline. While the hotel family scurries around getting boiled water and bedclothes, and trying to fix the water and the lights, we happily hang out the damp laundry. Our immediate needs, indeed all our needs, are met by the not-so-bad and functioning communal bathrooms, showers and washing sinks down the hall. Our bathroom is eventually triumphantly turned over to us but the floor is awash and the toilet still will not flush. Unused, it is a monument to the foreigner's (my) bad judgment!

I sponge myself off in what turns out to be the men's washroom (and toilet) and Kailin starts to bother me about my laundry again. I have missed the opportunity of the Taibai and most of what I have brought is rapidly reaching unacceptable levels of filth. Out of friendship, she insists. I hand over my two cotton blouses and join her in the (women's this time) washroom, working on my socks and underwear. So, temporarily, the subject is dropped.

As usual Robert is admiring the view, but Zhongling is worried and preoccupied. Supper is to be in the appalling dining room. I again suggest we try our luck in town. Our friends will have none of this. The problem of certificates has again come up. This time, we are told, the girl at the desk has asked for "protection money": ¥10 apiece. Zhongling has allowed her to understand that we will pay, if we last the night! Robert and I do not take it very seriously.

Supper is very good, belying the surroundings and the grimy persons of the waiter and cook. Yunyang County undisputed takes the prize for variety, imagination and best cooking on our trip, perhaps even including

Chongqing. Our meal includes fish-flavoured pork with Chinese parsley (cilantro), deep-fried fish, rabbit stew, and our favourite, egg and tomato soup. Kailin introduces us to bright pink slices of ginger, pickled in salt water to a sharp crispness.

The "protection money" is the main topic of discussion. On checking us in, the clerk had begun reasonably enough by asking a ¥2 service charge (each) for buying our boat tickets for tomorrow. She told Zhongling that we would have to buy third-class tickets, because of the demand for cheaper tickets by the local people, and that would be five times 20, ¥100 plus ¥10. I said that was reasonable.

Then, she came up with the ¥10 to guarantee our safety in this dangerous and out-of-the way area. (Only for foreigners? I can't remember.) Robert and I laugh. A foreigner's life is worth US$2.00! Zhongling explains that if we do not pay, the desk clerk will report us to the police and they will charge us much more. We joke that if it is really that dangerous (Yunyang seemed a small and amiable town and we have no intention of venturing outside in the dark), then we demand a Peoples' Liberation Army man, armed, to sit outside our door all night!

We will report all this to the Party Secretary, I grandly declare. Our Chinese friends just laugh. These counties are run by little emperors, and the little emperor IS the Party Secretary. The Communist Party, like everything else, has been told to pay its way in the socialist market economy. But Kailin declares herself not to be afraid (although she seems to think that Zhongling has handled the situation well). She says that if foreigners must pay, there must be a document to that effect. Show her the document and, even if it says ¥100, she (we) will most certainly pay. It is shameful that "the common people," a category in which she includes herself, are treated so badly.

I do not like the idea of paying protection money, but the sum is very small, and it is left at that. Zhongling is about to go to the desk with the ¥110 for the boat tickets when we conceive the idea of paying for them only upon delivery. No one thus far has absconded with our ticket money, but the amounts have been trivial. In a town where we are trying to avoid the reportedly not-so-good offices of the police, to whom would we complain? This is, it turns out, carrying bravura beyond reason, but our desk clerk, reportedly, agrees and so, we think, everything is in hand.

We leave the dingy dining room for the magic of the night on our balcony. The crickets are singing. Gradually the TV's with their Gong-fu shows are turned off. People sit outside talking, in the loud and harsh dialect of Sichuan. Otherwise, it is quiet. There are no policemen, no PLA men, no party secretaries. The lights reflected on the river go out one by one until it is almost completely dark. Just in case, I put the money belt around my waist. Then, there is nothing but the crickets and the sound of our laundry, finally and truly drying!

"As Much About Coal As It Is About Water"

(Robert, Yunyang, September 21, 1992) We sit and lounge and rest in the
growing heat waiting for the boat that would take us across the river to the
Zhang Fei Temple. I busy myself trying to get some good pictures of a scene
that is becoming very familiar on our stops but is not at all easy to capture
on film. In many of the towns and villages we have passed there were stone
ramps running down into the waters of the Yangzi. Some of these run
straight down for as far as a couple of hundred metres while others, where
the slope is steeper, are arranged in switchbacks. At their tops are either
the mouths of coal shafts running directly into the seams in the mountainsides, or else coal storage yards where black masses of ore from scattered
pits are collected for transfer into barges on the river.

   Coal is one of the most significant elements in the Chinese economy.
China has one of the world's largest coal reserves. Everything from primary industry to heating houses relies on it. In many places along the river
I see people moulding coal dust, which must be mixed with some adhesive,
into cakes or briquettes about the size and shape of a large soup can. These
little cylinders have holes through them from top to bottom. They fuel
household cooking fires, thousands of them in every town. Coal production
has been subsidized by the state through much of the communist regime.
This has led to massive inefficiency in its use. In Japan it is said that coal,
almost all of which is imported, is used six times more efficiently. Industries and public utilities in China burn coal wastefully while road, rail and
water transportation infrastructures are strained in the moving of the
bulky fuel. I have heard and read about energy waste in China but now I
am seeing many first hand examples. In one hotel room there is an air
conditioner set in the wall below the window. It is going full blast. Above,
the window cannot be closed because there is no glass in the frames.

   But in these riverside towns along the Yangzi all this coal theory is
quite academic. In a couple of places only, I see cranes moving the coal.
Everywhere else it is being moved by sweat and muscle. This means literally hundreds of porters moving in ant-like swarms up and down the ramps.
Each man carries a bamboo pole over his shoulder. A wicker basket in a
sort of open-ended cone shape is suspended from each end. They move with
quick, short steps, their knees slightly bent. At the bottom of the ramps
narrow planks lead up to the barges. The baskets are dumped and the
porters begin their return to the top of the ramps in a seemingly endless
chain of labour. The grime of coal is everywhere.

   Similar scenes occur at salt-works and fertilizer factories, but the coal
tips are most rivetting in their immensity. The problem with photographing the operation is that my wide-angle lens isn't wide enough to capture
the whole picture. When I try, the workers can't really be seen against the
sweep of mountainside. At the same time I can't really get close enough to

any one line of workers to catch them as individuals. And here is a central perplexity of trying to comprehend the way things are done in China: The long view fails to grasp the specific human dimension, but the close-up misses the sheer magnitude of scale. Furthermore, I feel extremely ambivalent about trying to photograph the scene at all. This isn't glorious work or heroic work. This is just grinding, hard work. There is something invasive and patently disrespectful about my sensitivities in treating it as some sort of novelty or curiosity for the foreign tourist to snap and show the folks back home.

Yet I also feel that this is part of the pulse of life in this valley. This swarm of humanity, wrestling the treasure of energy out of these mountains two basketfuls at a time and sending it down the river to fuel China, is a vital part of the work culture that will be irrevocably affected by the changes to be wrought here by new engineering and the floodwaters. This is part, surely, of what I have come here to try to grasp.

"This fascinates you, doesn't it," Zhongling says, jarring me from behind my camera and bringing me up from the dark well of my thoughts. I have been looking out from between two boats, attempting to get as close as I can to the workers without being totally obvious. He has caught me off guard and I'm a little embarrassed.

"Yes it does fascinate me. You see, almost no one in my country works like this any more. Everything is done by big machines. This kind of labour is something we've forgotten; it's outside our experience."

"I haven't forgotten it," Zhongling says. "Remember I told you about how I didn't get to go to university, because of the Cultural Revolution." We walk together back toward Kailin who is guarding our odd and ever-growing collection of luggage. "I was sent out into the country by the Red Guards, because my grandfather was capitalist. I worked for five years in a coal mine . . . doing that," he motions up the side of the mountain. I realize then that I can never really understand. While I had been taking education and opportunity for granted in the comfortable Canada of the 1970s, hundreds of thousands of my contemporaries in China had had their lives irrevocably disrupted. No wonder Zhongling takes his daily bathing so seriously. He must feel sometimes that he can never get rid of the grit of coal dust. There is nothing I can say. No assurance of empathy I might offer can relieve the memory of that deprivation. I can only admire the strength and resilience of Zhongling and of China. Here is a man about my own age who might have been bitter but wasn't. This country that has been shattered many times doesn't stay shattered. It moves forward with the determination of those porters on the mountainside.

The irony is that Zhongling is now an official of the state coal company. His job, in fact, has to do with energy conservation. He tours factories looking for ways to help them increase efficiency and cut waste. Some critics of the Three Gorges hydroelectric project estimate that as much energy could be derived in China from a better use of coal as will be realized from the output of the power dam, and with greater emphasis on investment in conservation technology, there would be a greater creation of employment. In some ways then, the Three Gorges Project is as much about coal as it is about water.

The work gangs loading the barges are ordinary working people, probably peasants who have drifted into this town looking for work. But it is their work, in part, that gives these riverbank towns their form. The stone ramps leading down into the river are not accidental constructions. Hundreds of years of trial and error, of building and rebuilding have given them their present shape. They are in fact constructed at just the right angle to allow men to walk down safely with a load and to walk up without becoming exhausted. The ramps are also built at just the right slope to enable a man to pull one of the ubiquitous two-wheeled carts up with a reasonable load.

The same genius is evident in the sets of stairs that ascend the banks in front of every town. These have a slope and step-height that is perfectly suited to the size of their users and the way they carry loads. Many of the stairs have landings or level spots at intervals that allow just the right number of rests to get a porter to the top. Many of the porters, especially members of a minority people called the Tujia, are equipped with baskets that they sling on their backs (see picture on the cover). The complementary piece of equipment carried with these baskets is a stout wooden "T"-shaped assemblage. The porter can rest his or her load by propping the stick under the basket to take the weight. This means that time and effort can be saved in not having to take the load off the back. Some of the stone steps leading up into the towns have small round indentations cut in them that accommodate the bases of the "T" sticks to keep them from slipping.

So just as the patterns of agriculture give character to the countryside, so the patterns of industrial work and commerce give shape to the towns. The culture of the towns is expressed, therefore, not so much through buildings but in the shape that established work habits have imprinted in the evolved design. These traditional work patterns are not expressed in the angles of stairs and the slopes of ramps alone. When we arrive in Yunyang, there is a pedestrian traffic jam on the town's main stairway just as we get off the boat. There is construction under way on a new building. Ahead of us on those steps people are scurrying out of the way and a great deal of commotion is evident. I run up to investigate. Before I can get close to the congestion I hear an amazing sound. It is a deep, rhythmic and haunting song. It comes from among the mass of people above me on the stairs. When I get through the crowd, what I see is quite astounding.

Days before, we had seen concrete slabs being manufactured in yards in the valley near Zhongzhou, further up the river. These standard-sized slabs are about three metres long, 60cm wide and 15cm thick. Each one must weigh a couple of hundred kilos. It turns out that they are used as modular floor segments in new medium-rise buildings. Until this moment on the stairs in Yunyang, however, it has never really occurred to me how they are put into place. What I see on the steps is a crew of four men arranged in two pairs. Each pair has a strong wooden pole stretched between their shoulders. Suspended from the poles are chains, and held in the chains is one of these massive pieces of concrete. As the four-man team moves slowly, step by step, up the steep walkway, they sing their work song with the same rhythm as their careful footfalls. I watch as they turn off the stairs and over planks onto the third floor of the building under construction. Their chorus ends with the musical clatter of chains as the concrete slab is lowered into place.

In the next days I hear other variations of these work songs which I am told date back to the days when enormous gangs of men called trackers hauled big boats up the river against the legendary current. The songs and steps and ramps, are, of course, not picturesque artifacts for the gratification of visitors. They are practical coping strategies for people who must work under extremely difficult conditions for simple survival. If they are seen as bona fide cultural expressions, the issue is certainly not whether they ought to be "saved" from change if that change has benefits for the people who use it. If they pass from practical use like the sailors' capstan chanteys of the days of sail in my country, then it might be advisable to record and classify and study them. The Three Gorges Dam will drastically alter the traditional way of life that I see evidence of and the communities that have grown to support and express it. I wonder if any of the benefits promised by the dam will actually accrue to *these* people? Will larger ships and mechanical loading of coal simply put all the coal porters out of work? Will they be able to find new and more humane work? Will the new towns that are to be built be easier places to work in, or will they have the same back-breaking labour, without the built structures that capture centuries of accumulated wisdom in ways that facilitate the work?

Zhang Fei's Legacy

(Ruth, Yichang, October 30, 1992) When one talks of the history of the
Yangzi, it is primarily about the State of Chu and the Three Kingdoms, as
if nothing much else happened. I remembered the first time I heard about
Liu Bei, Zhang Fei, and Guan Yu. These were the three men who swore a
famous oath in a peach orchard. (A fourth joined them later). In the late
1950s, I was going to attend a summer camp at Lake Couchiching north of
Toronto, an annual affair with young people of Chinese origin like myself
from the U.S. and Canada. Billed as a YMCA camp, it was in effect a singles
event, young people looking for mates. I thought it was a good idea to teach
them something about Chinese history and a friend, Jim Lee, suggested the
Three Kingdoms.

I started to read a translation of Luo Guanzhong's book and was fasci-
nated by the brilliant tricks and military strategies. I learned about the
alliance. To the tune of an old Chinese folk song, I wrote:

> *In a peach orchard,*
> In case you haven't heard,
> They made a vow,
> Honoured even now,
> To live and die together.

We had a four-piece orchestra with Chinese instruments and cymbals
playing the music. We had the principals, with their faces sticking through
cut-out holes above painted Chinese costumes. Unfortunately, the actors
were too busy courting to rehearse and the performance was a disaster.

But I learned a lot about Chinese history and for some time after,
people with the surnames Liu, Lowe, Lou, Lu; Zhang, Chang, Chong,
Cheong; Quon, Kwan, Guan and Chiu, Chu, Jiu (Canadian spellings of the
same surnames) would mention that I had composed a song about their
ancestors. I learned that the bonds made in the Han dynasty still held. Even
today the four families have a club, the *Lung Kong Tin Yi* (family association), in one of Toronto's Chinatowns and probably in other Chinatowns of
the world. But I doubt if they would still die for each other. There they drink
tea, have programs for seniors and juniors, help immigrants and contribute
or borrow from a credit union.

One of my friends, Ying Hope, who is actually surnamed Lowe, even
said he objected to the building of a dam that would destroy his ancestral
land. But I feel he is in a minority. We found that people in China who were
to be moved were eager to do so. Obviously, the correct *fengshui* of ancestral graves was not enough incentive to stay; anything was better than the
poverty in which they were living.

Near the Sanyou (Three Travellers) Cave was an historic pavilion and a new giant Buddha. They had been built to mark the entrance to the most dangerous gorge, the Xiling. A statue of the Three Kingdoms, general Zhang Fei looked out toward the water. Zhang Fei was a pig butcher (according to Moss Robert's translation of Luo Guanzhong's book) who became a general and governor of Yichang and died in 219 AD at the age of 55. He was killed by his own men because they feared his violent temper and death threats. "Better him than us." His face is usually painted red and black in Chinese operas.

Yi-Jing: Something Old That's New Again

(Caroline, Yunyang, September 21, 1993) When we awake, Li has already gone to finalize arrangements for our visit across the river to Zhang Fei Temple. We are to meet her there. Kailin has been busy as well. She has befriended an aunt of the hotel manager, who has agreed to help us. Solemnly we are introduced. We wish her family luck in their new venture. The dam, as everyone knows, will bring a lot of foreign business. Our status has risen in her eyes from easy pickings, to public relations possibility. We agree to the ¥20 as a late checkout fee. They will look after our luggage and let us rest in our rooms until the boat sails. Auntie has further decided that fourth-class tickets are a possibility after all. Without trepidation we entrust her with ¥60, and she sets off.

The fastest route to the small ferry to the temple is straight down the stone steps. Yunyang is getting back to work on Monday morning. Homes, stores and small factories open onto the narrow passageway. We are brought to a halt by a sharp explosion. A man is making popped rice on the steps. He has been heating an iron apparatus, shaped like a bullet a half a metre long, on a coal fire burning in a basket right on the steps. The bullet opens and hot popped rice pours onto a bamboo tray. He loads the popper again and puts it on the fire, turning it constantly. We ask, via Kailin, how long it will take to pop. Five minutes. Too long. As ever, we are late.

The next place that draws us is an old-fashioned spinning and weaving factory. The women wear surgical masks and cotton dust is thick in the air. Kailin promises we will be back and can photograph it then: we have an appointment at the temple and then the boat! Of course, we return another way.

Hurry up and wait: the small ferry will not go for another 15 minutes. A woman in a bamboo shelter sells tickets for a few *fen*. A young mother waiting anxiously warns her toddler away from the water. Everyone else is thoroughly relaxed as people slowly gather.

The ferry comes. Planks are thrown down on shore, and we climb from one boat to another, until we are high enough for the last leg up into the ferry itself. Seated on the benches that line the hull of this old tub of a boat, we turn for a last picture of the men loading coal. But as with so many other wonderful opportunities, we cannot get it right quite fast enough. The ferry swings out and turns around, fighting its way back up against the current, straining to keep from being swept back down river.

We are let off on the rocks at the steps to Zhang Fei Temple. The bank is marked with water elevation numbers. Women, selling bolts of red cloth, tapered candles, incense, and mounds of firecrackers, have set their Tujia baskets on the broad stone steps. Kailin and I chat with them. We have great good luck, they tell us. Tomorrow? The day after? Zhang Fei's birthday is very soon, and many visitors will come.

Kailin buys a candle and three joss sticks for Zhang Fei: a meager purchase, because intellectuals do not really believe. She turns to read the inscriptions on the stone walls below the temple. Some are just "from a person from Shandong" but there, above, is the famous: "Let us unite as one, against the great gusts of wind!" We take it as our motto.

The long steps turn at right angles, left and left again. Water gushes from the rocks above and falls to the river below. The flow is small this time of year, but our pamphlet shows the stream collecting into four pools as it descends the mountain. Isabella Bird found a "deep and narrow glen with fine trees and a waterfall, and over this a beautiful stone bridge has been thrown from the temple door" (1899:163). Although most of the temple Isabella Bird admired had been rebuilt only twenty years before her visit, worship on this spot must extend far back into Chinese prehistory.

A small altar marks the last turn of the steps. Pilgrims and petitioners burn incense and light their firecrackers here, to drive away evil spirits and to celebrate Zhang Fei's birthday. Terrific explosions and clouds of smoke bring us running to the top of the stairs. Only a wisp of smoke remains, and a ragged clump of torn red paper blowing away in the wind. A toddler methodically collects the bright wrappers: no one seems concerned he might find an unexploded firecracker among them. Kailin and I perch on the wide stone balustrade in the shade, waiting for the others, hoping someone will light some more firecrackers so I can record them on videotape. A ferry pulls up onto the rocks below us and people crowd off. There is a bustle of excitement as they stream into the temple, but no one has firecrackers.

Zhongling and Robert arrive, and then Li, and so our small group is reunited as one. Nie Shijia, the head of the local Bureau of Relics, awaits us in the temple's reception room. He graciously welcomes his Canadian visitors with gifts of sprigs of fresh longan grown on the temple grounds. These "Eyes of the Dragon" are like lichees; the rough husk must be peeled away and the tender and delicate white flesh is eaten from the large smooth brown pit. Tea is served. Mr. Nie apologizes that perhaps he and his colleagues will not be very skillful in explaining their projects to foreigners.

We ask the history of the temple and how it will be moved? "It is dedicated to Zhang Fei, a famous general just before the time of the Three

Kingdoms. For more than a thousand years it has attracted travelers because of its great beauty, and its stone engravings, calligraphy, wood carvings and paintings. Some, they say, come from the Han and the Northern and Southern dynasties. The tablets and other things are a way of teaching history to the common people."

"The foundations of the older higher levels date back at least to the Tang. *Zhu feng ge*, the pavilion for prayers for a fine wind and smooth sailing, was probably built by the Song. One tablet we will see describes the temple 800 years ago. The poet Du Fu tarried here for nine months; Su Dongpo, the famous Sichuanese poet of the Song Dynasty and the author of the two *Odes on Red Cliff*, sent poems and calligraphy from here to the court in Beijing, to be gifts for distinguished foreigners. "

"The lower levels, including the great hall, were swept away in the thousand-year flood of 1870 and rebuilt in the style of the Qing Dynasty. Today Japanese and South East Asians come to study the fine old things and to buy calligraphy and paintings. The proposed move will be just one of many in the evolution of the temple."

Yunyang will have a new county town. Where? The authorities have not decided. This site is under the protection of the Sichuan People's Government and they will oversee the move. Robert says he has read that the cost will be ¥4 million. Nie replies that is for the authorities to worry about. He and his staff are taking photographs and measurements of the valuable parts, the wooden beams. Beams, carvings, statues, the tablets, everything but the plaster and brick will be moved. If the temple were to be rebuilt from the beginning it would lose all historical value.

"This is not a temple, but an historic site. There are over a hundred places in Yunyang County, alone, connected to Zhang Fei, but this one is the most important." Zhang Fei is said to have drunk too much, beaten his soldiers and fallen asleep with his eyes open. In revenge, his soldiers killed him, cut off his head and threw it in the river. His eyes stayed open and that is why he is portrayed with large bulging eyes. To show respect for their dead general, the local people built the temple. (We can only suspect that local worship at the site much predated the perplexing hero Zhang Fei.)

There is, however, a nobler version of his death. In 223 AD, during his last and fatal campaign against the Kingdom of Wu, Liu Bei, ordered his friend and General, Zhang Fei, to lead his army from Langzhong County (northwest of Chongqing) to Jiangzhou (Chongqing). Two officers, Zhang Da and Fan Jiang, mutinied and killed him. The murderers, taking Zhang Fei's head with them, tried to escape and surrender to Wu. When they got to Yunyang County, they heard that Shu and Wu had made peace, and they threw Zhang Fei's head into the Yangzi. The head was retrieved by a fisherman and buried. Hence the saying that Zhang Fei's head is in Yunyang and his body in Langzhong.

"Are there any contemporary materials documenting Zhang Fei's life?" "A few. But we know he was a real person. Read the famous novel, *The Romance of the Three Kingdoms*; it tells the story."

In the third century, Chen Shou (238-300) edited the *Sanguo Zhi, The Collected Works of Zhu Geliang*. Then in 429, Pei Sonzhi, a court historian, revised and added to Chen's work. He did not feel constrained to produce an ethically pleasing or logical story, but went back to original sources, and compiled a document almost unrivaled in Chinese historical writing. Official histories of rival states, unofficial histories, biographies, clan registers and folk tales went into the revision, often as rival accounts of events in a very complicated period, that were even then passing into legend.

There are few historical references to these men between Pei's work and the writing of *The Romance of the Three Kingdoms* by Luo Guanzhong in the 15th century. The *Romance* is the *Iliad*, the Arthurian legends of the troubadours, Parsifal, and Shakespeare's history plays rolled into one, and just as politically motivated. And if the monk Gildas and the Venerable Bede, are western counterparts of Chen and Pei, then Geoffrey of Monmouth's *History of the Kings of Britain* (AD 1136) and Wolfram von Eschenbach's *Parzival* are to be compared with Luo's work. The novel, however, draws as much on an otherwise unrecorded oral tradition of poems, opera and puppet shadow plays as it does on the historians of the court. The heroes of the Peach Orchard, like King Arthur and Lancelot, lost the war but they won the peace. Ever after, story and song have favored Liu Bei, Zhang Fei, Guan Yu and Zhu Geliang, and their lost cause.

"In 1987, on Zhang Fei's birthday," Nie continues, "more than 10,000 people visited the temple in one day." (We are glad we are going to miss that, although it must be quite a sight.) We are constantly interrupted by blasts from ships' horns, and yet the temple never seems crowded. Nie is anxious to get away now because he has other guests.

The ground floor of the temple is a hodgepodge of passageways and courtyards, filled with potted plants, lanterns and general clutter. Robert and I dutifully photograph the carved and brightly painted wooden beams and rafters and pillars: the pieces that will be carefully reassembled on the new site. The Great Hall, "Friendship Tower," has been built as a backdrop for a tableau of plaster (Kailin would say "ceramic") lifesize statues of the three friends, Zhang Fei, Liu Bei and Guan Yu, swearing to live and die together, in a replic of the porch in the famous Peach Orchard (we assume on this very spot).

The three friends could only be characterized at this point in their careers as adventurers, but they lived in an age of social dissolution and that gave them their chance. Guan Yu (b. 162 AD) began adult life as a seller of bean-curd. He killed a magistrate, who forcibly tried to take the daughter of an elderly couple as a concubine, and fled. In 191 AD, he challenged to a fight a pig butcher, Zhang Fei (also b. 162), who had hidden his meat in a well sealed by a heavy stone. They were separated by a peddler of straw shoes named Liu Bei. Liu Bei claimed to be descended from King Jing of the Former Han Dynasty, of 200 BC, a tenuous claim to nobility. The three men became firm friends. (Another story has them meeting in a village inn, drinking wine.)

In 1594, Emperor Wanli proclaimed Guan Yu the God of War and thousands of temples were built in his honour. The choleric Zhang Fei became, what else, the patron saint of butchers. Perhaps the people loved him, no matter what he did, or because of it. But the ideal of eternal friendship celebrated here, like King Arthur's code of the round table, was their strength and their undoing. And it ensured their eternal fame.

Kailin bursts in on us. Come upstairs!

The second floor can only be reached by two narrow stairways, one at each end of the main hall. Where Buddha might be, there is a colossal seated statue of a demon, not a man, in full battle gear including bronze armor. His black face, his burning bulging eyes and his ferocious grimace are truly terrible. Offerings of red cloth offerings drape his shoulders. A high picket railing, stretching across this long and narrow room, serves as an barrier between the statue of the god and his worshipers. Burning joss sticks and candles have been set into a low trough-shaped stone altar. In front kneels a throng of supplicants and worshipers.

A woman in peasant garb kowtows before the statue in Buddhist fashion, and then looks up at Zhang Fei. She shakes and throws two wooden leaf-shaped dice onto the floor in front of her, again and again, keening, imploring Zhang Fei in a loud voice. She stops and lights a whole package of incense sticks from those already burning. There is little natural light in the room, and the air billows with the smoke and the smell of the incense. Kailin says red cloth and incense attract the good wishes of the god.

Lined up on the floor are rows of green and brown bottles. High up on the platform where Zhang Fei sits are the traditional gifts of fruit, peanuts and wine. A chicken with its throat slit has been placed between his feet, a gift for Zhang Fei. Blood is red, red will keep evil at bay. I have visited many Buddhist temples but only at the time of Spring Festival, in Dragon Gate Village near Kunming, have I seen the practice of folk religion.

Kailin translates. The woman's story is pedestrian. Her son is leaving for a job away from his home town. She is praying for his success. If you are having problems in your job, in your marriage, something evil may be interfering in your life. You can pray for good luck and good fortune to drive evil away. The dice are part of the Daoist practice of Yi-Jing. They have two faces: one convex, the other flat and grooved. You throw them six times, keeping track of the patterns. Then you look the patterns up in the ancient Book of Changes and it will tell you what to do, how to have good luck.

The bottles are full of vegetable oil. To achieve longevity, you take the oil into your mouth and then spit it into the bottle. The woman picks up her bottle and pours some of the oil into a dish and lights it. It burns with a slow flame, slow for a long life. Then she puts the bottle beside the others on a stand behind the rail.

Yi-Jing is something old that is new again in China. In fact, it seems to be replacing Qigong, if not in the lives, then in the small talk of our friends. As Kailin describes it: Chinese intellectuals discovered that it is very popular in Japan and in the West. If the foreigners like it, there might be something to it. She laughs and mentions some of our mutual friends who have taken it up. It is harmless, a pseudo-science, fun and yet it may do some good. The Yi-Jing abounds with positive thinking and folk wisdom. But Kailin and her friends' experiences are not those of the peasant women on the floor kowtowing before the statue of Zhang Fei.

The first time she saw these practices (she calls them superstition) was when her group of young people was sent to the countryside during the Cultural Revolution to learn from the peasants. The farmers burnt incense and kowtowed and tried, thus, to stop the Red Guards from coming! The Party, she believes, was right to ban these practices as a means of financial and political oppression of simple people. The implication is that intellectuals do not take these old things seriously, and so it is not important to worry about what they do. As the new reforms strip the government of control over the everyday lives of ordinary Chinese people, these old customs are coming up from underground.

Although this may not be a religious centre in the records of the Sichuan Bureau of Relics, it clearly is a temple for these worshipers. The statue of Zhang Fei here is light years away from the white-robed aristocrat on the gaily painted pavilion below.

Mr. Nie rejoins us at the Zhufengge (Helpful Wind Pavilion). We are in search of ancient architecture, but it is clearly the current temple that is his delight. "The lower garden is characteristic of Suzhou; there are small and delicate things in it, full of grace and elegance. The north has a rougher, tougher style. That is the style of the higher levels."

The sun is merciless overhead and yet the pavilions and walkways are cool: the architects of Zhang Fei's temple designed them to take advantage of the wind the god is supposed to command. "Boatmen and travelers made offerings, with firecrackers. If in haste, the captain would stop the ship in midstream. But it was better to come to the temple and kowtow. The god could help you sail smoothly for the next 30 li (15 km)." Firecrackers are used in China to celebrate marriages, welcome guests and to send off the dead. (When Norman and I arrived at SISU and were just going to bed, they lit an enormous string of fire workers outside the window. We didn't know if we were supposed to come out to acknowledge them or not.) I could not help thinking that moving the temple 30 km from this spot would destroy its historic function entirely. Who will come to pray for a fair wind?

Nie is proud of the nearby pavilion that shelters brightly painted bronze statues of Buddha and his disciples, gathered into his temple over the centuries for safekeeping. Preserving Zhang Fei Temple is part of preserving local history. And then he is gone again. Kailin and I reenter Zhang Fei's shrine on our way back down. Kailin lights our joss sticks and I put FEC¥50 into a box for donations for the reconstruction of the temple. Kailin says I will be very lucky today.

We meet back in the reception room and Nie agrees to join us for lunch. He suggests the Yunnian Hotel, the one I so unwisely turned down. Its bright and cheerful dining room on the second floor is packed. If Yunyang County has the best food in the area, surely the Yunnian Restaurant has the best chef. Nie has been here many times and knows just what to order. Small bean curd cubes enclose a ball of spiced pork, swimming in a broth of black mushrooms in a pottery pot. Steamed ground mutton, a local specialty often detested for its smell by Han Chinese, but mixed here with glutinous rice, a trace of *hua jiao*, and a generous garnish of chives and Chinese parsley, cilantro. The Moslem detests the Han pig; the Han the Moslem's goats and sheep. If there is a significant Moslem population here, we did not see it.

The Three Delicacies, including sea cucumber (slug, to be more precise) and two unidentifiables from the sea, are a popular treat all over China. That the ingredients are available everywhere is a testimony to the new prosperity. The *piece de resistance*, the *shuei mizi*, the special fish from the Yangzi, is only ¥20, a steal according to the ever practical Kailin. Much less than half the cost up river (and twice the cost of the next most expensive dish!). This scaleless fish, to me a catfish or a close relative, is a great and precious gourmet treat. It reportedly lives on the bottom of the river, not a recommendation to foreigners considering what goes into that river. It is spectacularly good, just barely steamed in a broth with many succulent stewed garlic. It is, as the Chinese like to say, "fresh and delicate" in taste and texture. No other words will do.

Our discussion of things historical continues over lunch, but not very seriously. The dining room is full. Lunch, but not power lunching as we would recognize it, is a serious business occasion in China, and a serious drain on state coffers, some say an abuse of the system. But in a society with few public entertainments, cramped housing and very little differential in wages, it is a way of rewarding employees and having a good time. The government tried to cut back on extravagant lunches by ruling that they must be limited to four dishes; this merely created a new market for very large platters.

Clearly, the noisy party of big drinkers at the next table, by now stripped to their trousers and brilliant red undershirts, are not expected to go back to the office and do any work. What is the bed in your office for if not for sleeping off lunch! The remains of twenty or more dishes are piled high on the table, destined for the slops bucket. Everyone is playing the ubiquitous drinking game: one, two, three. The person who is "it" displays one finger, two? The last to match him must drain his glass. They rise for toast after toast. The proprietor, who makes his easiest money from booze, joins the shouting. Anyone who thinks the Chinese are a reserved and quiet nation of small eaters and moderate drinkers has not been to one of these boisterous everyday workaday feasts.

Nie offers to go to the Lanxian Hotel to be sure that there is no more trouble. On the way, he explains that this part of Sichuan is very poor and that the police sometimes object to foreigners taking pictures of poverty. We should respect people's feelings. That is good advice, but we have just not seen what one sees in Africa or South America. We pay the final ¥20, no receipt. The obliging aunt has bought us tickets for the boat at 5:00 p.m.

But now Nie exacts his price. Li has agreed to look at some things the Yunyang Relics Bureau has in storage, to see if she can help identify them. She will rejoin us tomorrow about noon at Baidicheng. The ever frugal Zhongling goes to turn her ticket in. (Why, I wonder, don't we just go each time and get our own tickets?)

Our final words to Nie are that there might be an international campaign to help save the rich legacy of Zhang Fei Temple. Perhaps, Mr Nie says, but he is only a local office worker, who knows nothing about these things. In earlier days, the expectation would have been British sterling or Yankee gold. Today, most offers of help come from the Japanese, but often with conditions that are unacceptable.

Our rooms have been straightened up. We stretch out for a much appreciated *xiaoxi*. And so we harbour no hard feelings toward "the best hotel in Yunyang."

Ancient Architecture And "Fake" Temples

(Kailin, Yunyang, September 21, 1992) With each and every passing day, things in cultural sites change their colours, paint peels off, they even change their shapes. They may be damaged, or even wrecked and not handed down from past generations. To protect our legacy, and to pass it on, we must repair or copy relics, and rebuild cultural sites.

Because of the construction of the Three Gorges Water Conservancy Project, many temples along the Yangzi River will be rebuilt. People will naturally worry that, in the reconstruction, the "real" sense of the temples will be lost. As for the charge that they will be "false" temples after reconstruction, I think that is not wholly true.

During the period of reconstruction, the Zhang Fei Temple, for example, will be moved to a new site. According to Mr. Nie, the curator of the Cultural Centre of the Zhang Fei Temple, the whole wooden part of the structure will be moved. So, basically, the original structure will be intact. It will not be entirely the "false temple" that people imagine. The decoration and the touchup of the reconstruction will mean that the new site will not only keep the quintessence of the temple but will increase its graceful bearing. This move and reconstruction will add a new chapter to the history of the cultural site. In this way, the sense of loss will be balanced by a fresh new vision and a new page will be turned for tourism in the Three Gorges. (We have to remember that there has been tourism here for thousands of years!)

However, some things can't be reused or reproduced; there are special land forms and scenery that cannot be moved. There is no alternative here to accepting the "false" and to finding a remedy or a way of making up for what is lost.

Moving Zhang Fei Temple

(Robert, Yunyang, September 21, 1992) The compound perched above the river opposite the town of Yunyang is clearly the largest and most important single complex of historical architecture that will be directly affected by the flooding of the Three Gorges Dam reservoir. It is known by at least three different names. The first of these, Temple of the Ethereal Bell of Ten Thousand Ages, pushes the limit of translating colourful Chinese phrases. Willow Prince Temple and the title most commonly in use now, the Zhang Fei Temple, are somewhat more prosaic.

The *fengshui*, or auspicious placing of the temple in terms of surrounding conditions and land forms, seems to be ideal. There is a large mountain behind, smaller ones opposite and a waterfall issuing from the very rocks over which the buildings have been built. The temple faces north rather than the prefered southern orientation but that aspect of *fengshui* seems to be less of a factor in the rugged conditions of the Sichuan mountains than in the plains of northern China.

The structures of the Zhang Fei Temple themselves are not among the most intricate or elaborate examples of devotional architecture in China, by any means. They are in fact quite modest, and differ from the rural buildings we see more in scale and in decorative detail than in their basic construction. They consist of massive timber pillars supporting equally massive roof beams. The largest of these are about 40cm in diameter. The roof beams are configured in the characteristic right-angled arrangement, joined by cross-members and held in place by pinned tenons in mortises. Over the purlins supported on these frames I can see rafters, and then plaster. There may also be planking between the rafters and the plaster as indicated in some literature on Chinese construction methods, but it is not visible from my inspection. Unlike any of the rural buildings I've seen, the roofing plaster is covered not with black clay tiles but with brightly coloured ceramic ones.

The upturns on the ends of the eaves are much more pronounced than on other buildings I've seen. That is apparently a feature of more southern Chinese architecture. Mr. Nie, the curator of the site and our host, explains that this complex was a mixture of styles and represented a meeting of northern and southern architectural influences. Other differences from buildings in the countryside are the complex wooden lattice screens between rooms, decorative brackets on the pillars and intricate inlaid coffers in the ceilings. The decorative and lattice work was quite fragile, consisting of wood only about two centimetres thick. Besides the wooden lattice room dividers there are some screen walls that are probably of plastered brick. I get a better look than before at the way in which the doors in this kind of building are set, top and bottom, on pintle hinges formed by projections carved out of the surrounding masonry. As well as rectangular doorways that can be closed there are also many examples of round openings between rooms. The entire building is set on a flagstone base, and the individual pillars are supported on stone drums only slightly larger in diameter than the pillars themselves.

The temple buildings are on at least two main levels. The older part is further up the slope. The sections we enter first were rebuilt about 120 years ago after a flood carried off many of the buildings closer to the river. On the outer wall of a gallery on the lower level we are shown a painted inscription. The number 156 is prominently displayed. This marks the high-water level of the 1981 flood, 156m above sea level. Since the planned level of the Three Gorges Dam reservoir will be 175m above sea level, it is graphically clear that the whole of the Zhang Fei Temple site will be submerged. The stated plan is to move the complex. Aside from the questions about the traditional placing of the temple, its relation to the river and its function as a devotional site, there are many serious considerations arising from the proposal to relocate it.

There are at least some previous examples of historic buildings in China being moved to facilitate other civil engineering projects. The Yongle Palace is a group of 13th century AD buildings that originally stood in Yongle City on the Yellow River in Shanxi Province. It was relocated to Ruicheng to make way for the Senmen Gorge hydrological project well over ten years ago. In that case, Yuan Dynasty wall paintings were preserved, and I assume the original upper timber work, and perhaps even the roof tiles of the buildings, were also moved. When the site of the Qu Yuan shrine near Zigui on the Yangzi was to be flooded by the reservoir of the Gezhouba Dam in the mid 1980s, however, the original building was not moved. Instead, a not-very-accurate replica was built. But the Chinese have many more pressing problems to deal with concerning projects such as the Three Gorges Dam. It is in this regard that the assistance and financial support of Western architectural preservation experts might be of some help.

Most such work is guided by the terms of what is called the *Interna-*

tional Charter for the Conservation and Restoration of Monuments and Sites. More commonly known as the *Venice Charter*, this undertaking was developed in 1964 by a conference of the United Nations Educational, Scientific and Cultural Organization (UNESCO) and at least nominally adopted by most member states of the UN. I'm not sure if China is a signatory. The *Charter*'s main points deal with the definition of an historic site or monument, with conservation and with restoration. Historic sites and monuments are defined as examples of buildings or settings in which can be found evidence of a particular civilization. These might be great or humble structures, but ones which have acquired cultural significance with the passing of time. In safeguarding such places, those responsible ought to have access to the best scientific and technical advice available. In the conservation of an historic monument, a setting should be preserved which is in scale with the original. Moving all or part of a monument ought not to be allowed except where it is the only way to preserve it. In restoration the aim is to preserve and reveal the aesthetic and historic value of the place. The contributions of all previous periods ought to be retained since a unity of style is not as important as an appreciation of the ongoing significance of the site. Restoration should be based on authentic documentation and ought to stop where conjecture begins. There should be clear indications to a viewer what parts are original and what elements are restored. And finally, modern techniques should be used only where traditional ones are inadequate for the task.

The principles of the *Venice Charter* probably apply more appropriately to considerations involved in moving the Zhang Fei Temple than to any of the other sites we will see. There are few other places that constitute such specific and definable expressions of material culture. The temple complex, while not a unique or superlative example of Chinese architecture, certainly has acquired cultural significance with the passing of time. Besides being significant in themselves, such temples have also served as regional repositories for religious articles from other sites. This one contains a collection of inscriptions, calligraphy and devotional objects such as statues that people have brought here for safe-keeping over the centuries. The temple's conservation and restoration certainly deserves the best available scientific and technical expertise. But that raises the larger matter concerning the allocation of financial resources to the conservation effort. One document relating to the Three Gorges project suggests ¥4 million (US$800,000 at 5:1). I would have to say that if anything close to Western standards of conservation practice are followed, that estimate is dangerously low. The question of moving the temple, however, is beyond debate since failing to do so would be to consign it to destruction. What needs to be considered is not whether, but how it is to be done.

The first point would be the appropriateness of the new site. It should duplicate both the lay of the land of the original and the overview and proximity to the river. There should be little difficulty in moving the main vertical and horizontal timbers since they appeared to be in good condition. Testing them for soundness, however, might be an early priority. The problems that I perceive are with the decorative woodwork and ceramic roof tiles. Much of the wooden detail work is in such poor condition that moving it will either be extremely expensive, since it will need to be carefully removed and restored, or it will be impossible. Many of the inlaid ceiling coffers are in even worse shape. Much of that work will need to be replaced and the question of finding qualified craftspeople will arise, although I understand that much traditional woodworking is still done in China. The colourful tiles are also in poor condition, and removing them from their bed of plaster will be difficult.

One of the ideas expressed by Mr. Nie, the site curator, is that just the upper wooden components would be moved. That makes me wonder what will happen to the stone foundation work which is obviously an integral part of the complex and probably predates the wood. I also ponder the fate of the roof tiles if what he said was true. Will there be any excavation of the foundations and surrounding area?

Perhaps the most difficult problem of all, however, is documentation. I have very little idea of what kind of historical material is available to assist in accurate restoration. On the other hand, I'm not sure that documenting the accuracy of specific historic details is as much a priority in the Chinese view of historic sites as it is in the West. I got some feeling for the different way of viewing things when I returned to Canada: I met a graduate student who has recently emigrated from Hong Kong. His family name is Cheung. He reminds me that he is a member of one of the families that consider themselves descendants of Zhang Fei. Some of the others are Chung, Chong, Cheng and, of course, Zhang; in other words, all the people with similar-sounding family names. Whether modern Chinese people take these matters of descent very seriously or not, he is certainly quite familiar with the legends of his supposed ancestor. When I mention the concern about moving the temple away from its original site and possibly even away from the river where Zhang Fei's spirit is invoked, he laughs. But it is not because he thinks the ideas about good luck on the river are unworthy superstitions.

"Don't worry," he says, "Zhang Fei had such a loud voice that he will be able to be heard no matter how far he is from the Yangzi. Once in a battle he shouted so loud that he caused a bridge to fall down and that was how he defeated his enemies."

So perhaps the Western preoccupation with accuracy and documentation is not necessarily considered important in the Chinese milieu. Still, there is an opportunity presented with the relocation of this temple. A baseline of documentation could be established in which the best estimates for the construction dates of various periods in the complex could be established. These dates could be arrived at through the application of such techniques as dendrochronology (the study of tree rings), chemical analysis of paints and other elements, carbon dating and other methods. Excavation of the foundations and surrounding area could also yield data to help establish the chronology of building on the site.

If foreigners are to be helpful preserving this historic site then a strategy for approaching and executing the move will have to be devised that is practical and adheres as much as possible to high standards of heritage professionalism. It must, however, in the final analysis be a made-in-China strategy. The approach must suit the local cultural situation and, while it may draw on foreign ideas, it must in no way be imposed. Foreigners must remember that if they sincerely want to help then they should be assisting, and not presuming to lead.

7

FENGJIE

The Yellow Snake

(Caroline, on to Fengjie, September 21, 1992) Safely on board Boat 113, in our fourth-class cabin, we recap the successes of the day, and the spontaneous kindnesses of people who do not expect to see us again. The fears of the police of the night before, which we hope are only holdovers of the past in the minds of our friends, are almost forgotten. The wind comes up. Zhang Fei's spirit cools us, I think, and the river narrows. The sun sets gloriously behind us, and the PA system plays Chinese Country and Western music. Fishermen and aggregate miners take shelter behind the spits of rock. Travel on the river, travel itself, has established a rhythm so that, although we have only been gone a few days, we feel suspended in time.

I venture out into the wind and sit on the capstan where I can see the shoals, and the whirlpools. The wind and the waves constantly change direction around the headlands. In the wake of our boat, pink water blends with pink sky, as the sun sets. There are not many farmhouses or terraces here. A few pines straggle up the barren rock. These mountains have heaved themselves up, their strata almost perpendicular. A father and his bored six-year-old play with a yellow plastic airplane. Robert calls us to come and see a sampan with a sail, the only one we have seen. We steer into the waves and even our stolid ferry bounces and thuds. We pass Ferry 61, a small one. These passenger boats vary greatly in size. The sun has almost disappeared and the mountainsides and river are very blue. A woman approaches selling boiled eggs and oranges in a string bag.

This is the only ferry we will ride that is not packed to the limit. Even the cafeteria is quiet. After our huge lunch we will wait to eat at Fengjie. The music has stopped, the light has disappeared. We return to the cabin and Kailin offers to translate one of the legends from the book she has bought.

The Three Gorges area is famous for more than the legends of the

Three Kingdoms. In Chinese fashion, Kailin categorizes them. First, there are the nature legends, of mountains, streams and springs, of caves, mysterious rocks, strange scenery, and precious and rare flowers and trees. Then there are legends of villages, pagodas and temples, cliff burials, plank roads, generals, and of princes and princesses.

Pick one, we say. Kailin picks *The Yellow Snake who wanted to become a Dragon.* "In ancient times, it was quite possible for an animal to change itself into another, if it trained faithfully and followed the correct procedure [rather like becoming a mandarin or a party secretary!]. But becoming a dragon was very difficult, for a dragon is powerful and leads a happy life. The Yellow Snake was lazy and impatient and sought to shorten its path by presenting treasures to the Dragon King. At last, Dragon King acknowledged these presents and asked Yellow Snake if he had any requests. The wily snake said only, 'Introduce me to *Guanyin*.' The Dragon King said this was easily done. Soon there would be a meeting of fairies from all over the world and Guanyin would be there."

Guanyin is the Goddess of Mercy and of Children. Every area of China had its female deities and these blended into the ideal of Guanyin. Eventually the age-old Guanyin was identified with a female disciple of Buddha, *Avalokitesvara*, just as Christianity incorporated the worship of local goddesses into the Virgin Mary. Pottery statues of Guanyin, sometimes with a baby in her arms and looking quite like a Chinese Virgin Mary, can be bought in any town in China today.

"Is a Chinese fairy like a leprechaun or like Tinkerbell?" we ask. "Some are male, some female: they look like people but they wear old-fashioned clothes like the characters in opera and they can fly and work magic. Fairies can do good deeds for human beings, and so we seek their help." There is a famous story about Liao Zhai, a fairy who became a human being for love of a man.

"Yellow Snake went to the fairy meeting. As promised, Dragon King introduced him to Guanyin. Yellow Snake kowtowed and praised himself. But Guanyin was impatient, for she had seen the likes of Yellow Snake before. She agreed to give him a golden potion a thousand years old if . . . ."

The banks are very close now. It is almost pitch-black dark and the ferry's spotlight sweeps the rocks, back and forth. Kailin is tired of translating and she complains that I set too many tasks! Guanyin finally loses patience with Yellow Snake. He still has 52 years of testing to go. "You can never be a Dragon; you will always be a yellow eel!" We rest.

Fengjie is a major coal centre. Mine shafts seem to have been dug straight into the rocky banks of the river. The miners simply dump the laden ore carts down the slope and dock workers shovel it into baskets and carry it onto the barges. The ferry must wait ten minutes before it can dock. All boats stop here for the night, entering the first gorge only in daylight, and so the facilities are very busy.

Fengjie has the longest set of steps yet. There is no way I will make it with my share of the gear. This town, however, is different. There is no milling crowd hustling business, but a man with a megaphone assigning porters. A much bigger man than yesterday presents himself. Zhongling suggests ¥2. We have no idea of the distance to the "best" hotel. The man protests it is far, and it is. Zhongling agrees to ¥4, drops his bags and starts grabbing ours.

Our worry about loading the man down with too much soon changes to worry that he will disappear far ahead of us (we never relinquish the cameras). But even the porter is forced to stop half way by these daunting steps. Kailin and Zhongling are directed to Guesthouse No. 2. It is a very long way but, once up those steps, the town is relatively flat and the streets are wide and well-lit. Many buildings are new and the sidewalks are alive with young people.

Finally we arrive at the hotel. Zhongling gives the porter ¥5 and he disappears happy. Robert and I gratefully sink down into the lobby chairs and Kailin and Zhongling start the negotiations. They do not go well. Guesthouse No. 2, despite Zhongling's protests and Kailin's effusive admiration of their many certificates and awards of excellence proudly displayed on the wall, is not allowed to house foreigners. We must go to Number 1, a "few minutes" away. A boy hanging around the lobby volunteers to be our porter and off we go. Not far is far. No supper yet.

The lobby of Guesthouse No. 1 is bleak, but a large archway leads into a garden. Perhaps this is a garden hotel? Those explosions are firecrackers welcoming us? The fierce rat-a-tat-tat is stones in an automatic cement mixer. A new wing of the hotel is under construction and will remain so until 11:30 p.m. As in Hong Kong or Shanghai, the workers are working under a huge artificial light.

We are welcomed by the local Party Secretary (or the party secretary of the hotel, we are not sure). He invites us into his office, which is also plastered with certificates and banners testifying to the excellence of this hostelry. The volley of stones stops, but, alas, begins again. The Party Secretary is a compatriot, a native of Chongqing. Kailin's antennae are activated. Perhaps we will have some *guanxi* here. We will need it. The water is not on and beside each toilet is a bucket. Even when the water is turned on, the toilets will not flush. The beds have mosquito nets, and they are not for rustic effect. The tubs are worn and stained brown, the floor wet. We ask for another room, hoping for a working toilet. Too soon it is clear that they are all the same, but at least we find one with a toilet seat.

It is 9:15 and we are far past the dining room supper time. Perhaps our host can direct us to a restaurant? I have been looking longingly across the street at a tiny hole-in-the-wall noodle shop, but Kailin is not convinced it is safe. Our host leads us to the special second floor of a private restaurant around the corner, where he seems very much at home. Four gentlemen, now stripped to their undershirts, are quietly enjoying a hot pot in the corner. The restaurateur, our host's friend, offers fresh seafood and fried pineapple. He is dismayed that we want only noodles and *jiaozi* pork dumplings. We are bushed but we insist. Finally the teenage waitress brings us three steamers of *jiaozi*, with an excellent vinegar, soya, chili and garlic sauce. Eventually the noodles come too. The bill is an astounding ¥35. Kailin estimates the Party Secretary's commission is ¥10 but Robert and I don't care.

Our host is a great booster of his new home town and a fount of information on the future prospects of Fengjie and the Three Gorges Project. He is also enthusiastic about the socialist market system and invites us to join him in a joint venture, any joint venture. He has had American, Japanese, and even Canadian guests. We ask him when? Were these Canadians engineers making a study of the dam? Our friend is not telling. But the Canadians were prodigious drinkers.

He is so carried away that he offers to turn the hot water on in our rooms. When it eventually comes, there are no plugs for basin or tub, and it slips away despite my frantic attempts to jam my washcloth into the hole! Kailin insists on helping me. I submit: I have my journal to write. Finally, near midnight, the noisy activity of the courtyard ceases. The workers go home, we hope, for bed.

City Walls and River Gates

(Robert, Fengjie, Sept. 22, 1992) Amid the confusion of the morning that eventually sees us paying the local Communist Party jeep and driver to disappear, Caroline and I have a mission. We want to see and photograph the site known as the Fengbi Wall. It is only a few blocks away from the hotel, but they are blocks of narrow streets crowded with the tables of free-market vendors. The difficult part is keeping Kailin and Caroline from stopping to shop. Kailin explains that many of the people running these stalls probably have regular jobs elsewhere but this is their transition to a competitive economy. Their housing and benefits and a salary come from the factory or office to which they have been officially assigned, but they make their real money selling in the streets. There is also a bewildering array of foodstuffs on offer in these crowded lanes — peppers and spices and things that look totally exotic to me. There is an excitement and energy in the air. Kailin is in her element. She explains as much as she can of what we are seeing, but I keep urging us on to the "historic site" that is the objective of the walkabout.

Walls have played an extremely important part in Chinese history. There is a wonderfully concise and insightful little book called *The Art of War*, which was written in China in the fourth century BC by Sun Tzu. I read it as an undergraduate in university when I studied military history and more recently it has sold thousands of copies in North America. It is acknowledged as the oldest-known treatise on military science, and its principles are as applicable today as when they were written. Notwithstanding the fact that the Chinese developed very sophisticated siege equipment, Sun Tzu cautions against attacking the enemy's cities. He explains how long the preparations to overcome the walls will take and estimates that one-third of an attacking force will be killed if they assault walled cities directly. It is clear from such passages, as well as from archaeological evidence, that Chinese cities were massively defended from the earliest times. There are even traces of moated and walled settlements from the Neolithic period.

The most ancient rules of Chinese town planning, the *Zhouli* or *Zhou Rituals*, stated that the first thing that was to be done in building a city, once the auspicious site had been selected, was to erect the walls. From very early times these were constructed of tamped earth. Often they were of extraordinary dimensions. The Han capital at Changan (Xian), for example, was begun in 192 BC. Its walls were 25km around, 18m high and 15m thick. The walls of smaller cities and provincial towns were more modest of course but, nevertheless, all towns were walled. While the vast majority of the Chinese people have always lived, and continue to live, in undefended villages in the countryside, walled towns have been the centres of control. Behind the walls tribute grain and other valuable goods were stored. The wealthy, the powerful and the soldiers lived behind the walls. For much of China's history there has been a need to defend against the incessant barbarian incursions from the north. The Great Wall, after all, was an attempt to protect the whole country, but the walled towns also stood against the periodic social disorder, banditry and the chaos of dynastic succession and rebellion.

China's walls, however, have had an even more significant psychological role. They were, in a way, symbols and beacons of order. In traditional Chinese urban form it has not been particular buildings that have been considered paramount or focal in the way that a cathedral or citadel is in Europe. Rather communities have been treated as works of planning and architectural art. In that context it was the walls that defined the space and which were usually its largest components. The gates and corner towers were generally the tallest structures and were often bigger than the palaces in the capitals or the governors' residences in provincial towns. Those important buildings in China were not necessarily to be viewed as imposing edifices from a distance, the way a castle often dominates a European town, or bank towers rise in the centre of a modern North American city. They were approached through a series of gates in the internal walls within a city. Those internal walls were also governed by planning principles. The classic Chinese city was divided into walled districts called *fangs*. Each *fang* incorporated several blocks and there were as many as sixty in the great capital cities. Some of these were occupied by members of the same trade or guild and some were family compounds. All were locked at night. The modern extension of this tradition is the factory or work unit compound, walled just like their predecessors. Steel mill or university, the specific doesn't affect the form.

There exist many accounts, some from well into this century, of the walls of the towns we are visiting. In the days before arriving at Fengjie I had seen a number of massive stone structures on the sides of the river. These are the sorts of things I have come for, the kind of historic sites I know something about, architectural treasures, embodying the past that surely, must be saved from the rising waters of the Three Gorges Dam reservoir. The only problem is that, on closer examination, some of what I think are ancient forts turn out to be lime kilns. They are not old military architecture but industrial installations. They are probably only a few years old, turning local stone into the raw material of cement. Others of my "walls" and "towers" turn out to be loading platforms for factories or storage bins for coal mines.

Like just about everything else in China the once mighty walls, with all their strategic and social importance, are gone. In the larger cities the walls have sometimes been removed to make way for ring roads just as they were in many European cities such as Vienna. At various times in Chinese history the tamped earth ramparts may have been faced with brick or stone, but both those materials are eminently reusable and much in demand. In the smaller communities we visit it may well be that stones now embedded in the sides of this lime kiln or that factory foundation may once have come from nearby town walls. But the fact remains that once neglected and uncovered, even the stoutest of Chinese defences simply melt and disappear in the floods and seasonal rains.

That is why the Fengbi Wall is so important for us to see. Except for a structure in Chongqing, that is said to be the remnant of that city's once massive wall, and a couple of other minor examples we missed, this is the only piece of military architecture extant in the whole valley. Approaching it from the rear or inside, the effect is not overwhelming. The wall, at the end of one of Fengjie's crowded market streets, is only about six metres tall. In fact, it doesn't really look like a town wall at all. It looks, instead, like just the lower part of the buildings that have sprouted from its upper surface. Only the absence of windows and the arch of the gate, which is perhaps five metres high and eight metres wide, give it away.

When we reach it, the dimensions become more impressive. The wall obviously stands at the very top of the steep riverbank. Stairs lead down through the gate and reveal that the wall has been perhaps 12m thick. That is in keeping with the archaeological findings in the great capital cities. When we emerge on the outside of the gate, the wall is considerably more imposing, 12m in height. It appears that little more than the gate itself remains. On either side, the structure simply disappears into the surrounding buildings. Those structures may have been built of materials from the former wall or may have incorporated parts of it in their own fabric. Above the arch of the gate is a stone panel in which three characters have been carved. I recognize the first as the fairly pictographic character for *men* or gate. Kailin says the other two were the name of a poet, Tai Bai. Our dubious host from the night before, referred to this as the "Poet's Gate."

But what exactly is it that we are looking at here on this sunny morning at the top of another of the endless sets of stone steps that lead up from the river boats docked along the Yangzi? The Fengbi Wall was not the only or even the main gate of the old city. A writer from an earlier time has described Fengjie in this way:

> The town, which has a high wall, is on a steep slope bounded in four terraces and with a small pagoda behind. Facing the river is the imposing main gate with curved tiled roof to the double-storeyed pavilion above the gate. On the face of the wall above the gate itself is the inscription "Slow" written in vast letters. This refers not, as might be surmised, to the character of the inhabitants or the trade of the place, but to the speed at which it is desired ships shall pass the town so as to avoid inflicting damage from their wash to junks banked in. (Worchester 1971:498)

The Poet's Gate, then, is perhaps just a serendipitous remnant of a whole order of urban form both here in this one town of Fengjie and in the whole valley. This particular section of wall is stone faced and solid looking. But I see no indication of hardware that might once have held the closable gates themselves. This makes me wonder what degree of rebuilding has already taken place. What we discover as well is that while this town, no doubt with walls, may have existed for hundreds of years, the present Fengbi Wall probably dates only from the Qing or Manchu, the last dynasty, which came to an end in 1911. The wall we are looking at, then, was probably not more than a hundred years old. Did it, perhaps, never even have a gate, but was it simply a kind of ceremonial passageway into the town that recalled the earlier defensive works?

My understanding of the CYJV survey is that it recommends that the stone plaques with the inscriptions be relocated, but not the wall itself. We are not sure if that means just the panel above the gate with the three carved characters or whether there are other inscriptions. It seems to me that the Fengbi Wall, and the few other remaining fragments of town ramparts that exist in other communities, might reward some degree of archaeological investigation of the base and surrounding area.

It isn't that I didn't think that the Fengbi Wall, that one remaining fragment of such a monumental aspect of the area's heritage, shouldn't be commemorated in some way even if it were only preserving the stone tablets. That is one thing that can definitely be accomplished to save something of the past. It is just that I am not sure that sort of gesture makes much sense in Chinese culture. After all, the rest of that town's wall and all the other town walls have already gone, not because of a state-sponsored mega-project, but because the towns have continued, as they always will, to evolve. But I am not sure either that all the walls have totally disappeared. Perhaps if there were time and resources for proper studies we could learn more about the old walls. Perhaps some older people in these communities know where traces of the former structures are hidden among the later buildings. Their recollections are the more significant cultural heritage aspects that will be lost after the people of these towns are moved to newly constructed communities.

"The Wise Are Like Water"

(Kailin, Fengjie, September 22, 1992) Nature created the beautiful mountains and rivers of the Three Gorges. The two hands of the Chinese people have made the bright culture of the Three Gorges which enhances the beauty of its mountains and rivers. For several thousand years, the gorges have been one of the main scenic attractions in China, and the second cradle of ancient Chinese civilization.

The Three Gorges consist of the Xiling Gorge, with its many shoals and rapids, the deep and magnificent Wu Gorge, and the imposing and dangerously steep Qutang Gorge. Their magnificent scenery is comprised of four aspects. There are peaks, precipices, bends and shoals. Strange peaks stand lofty and towering into the clouds. The most famous are the Twelve Peaks of Mount Wu, Mount Tianzhu (Heaven's Pillar) and Sanbadao (The Three Swords). Sheer precipices and overhanging rocks line both sides of the river. Queer-looking rocks are named after the livers of oxen and the lungs of horses. Others are identified with books on the art of war, double-edged swords, a man leading an ox, and even a monk hanging by his feet. There are also carved poems and prose under overhanging rocks. The White-washed Wall Precipice in Qutang Gorge is extraordinarily magnificent.

The Yangzi snakes its way through the Three Gorges. Sometimes a big mountain ahead gives the illusion that the river comes to an abrupt end. But it winds around and surges onward. There are a good many shoals and rapids, where the waves are high. In every season, whether flood or low water, danger still lurks. The Kongling and the fierce Xintan shoals were the most notorious.

The Three Gorges have long been an important strategic area, the site of events of great historic interest. Men of letters and famous calligraphers have frequented the area since the earliest times and it is the birthplace of

Qu Yuan, the soul of China, and of Wang Zhaojun, who was as brave as she
was good and beautiful. Many forms of classical poetry, such as *Chuci* (The
Songs of Chu), *Hanfu* (a popular poetic style in the Han Dynasty), *Tangshi*
(poetry of the Tang dynasty), and *Songci* (poetry in Song Dynasty) have
found expression in verses about this area. Great writers, including Li Bai,
Du Fu, and Su Shi, wielded their brushes and left their wonderful characters for a hundred generations. Politicians and military leaders such as Zhu
Geliang, Guan Yu, Kou Zhun, and Yang Xiou ruled the populace or led their
armies though the Three Gorges. In their spare time, they too wrote poems.
Geographers, tourists, exiles and scholars also travelled here, for the mystery and for the wonder. There is an old Chinese saying, recorded in the
Lun Yu (The Analects of Confucius), that, "The wise are like water, the
benevolent like mountains."

Although the Three Gorges has long been a popular place for the Chinese to travel, it is now developing as an international tourist area. To allow
it to reach its potential, however, a lot of problems must be overcome. In
some places there are no hotels at all for foreigners. While some of the
hotels where we stayed were good, such as the Taibai Hotel in Wanxian,
others were not. The hotel in Fuling was the worst. When we arrived we
wanted to make some tea. Robert picked up a cup. There was a dead fly in
the cup and a note in English which said, "disinfected." "Look," Robert
said, "they've disinfected this fly for us." We all laughed, but if we want
more tourism in China we have to eliminate such jokes.

Practice proves that it is reasonable to face the world, and tourism can
become an important economic factor in that reality. Three Gorges tourism
must formulate a management policy to widen its scope. To achieve this,
not only the state tourism department should be involved, but also the
provincial authorities and right down to the basic work units. We must call
on the masses to do everything possible and expect every effort in becoming
masters of their own affairs. We should concentrate the necessary man-
power, material and financial resources and grasp the favourable opportunities using pictures, sound and writing to publicize the natural scenery and
places of historic interest. At Baidicheng we met a young girl who was a
guide. She wanted to learn English and came in the evening to talk to us
about how she might do that. None of the guides spoke any foreign languages. Young people like her should be encouraged to develop their skills
to help build the tourism industry.

8

BAIDICHENG

The Battle of the Red Cliffs

(Ruth, Yichang, October 29, 1992) My guide Susan took me in her company's minibus to the suburbs, to the entrance to the Xiling Gorge, the most easterly and dramatic of the Three Gorges. We peeked into the Cave of the Three Travellers, embellished with the poetry and white statues of the poets. Susan mentioned one of them, a Northern Song poet named Su Dongpo who wrote a famous poem about the battle of the Red Cliffs that took place a couple of hundred kilometres east, near Chibi, between Wuhan and Yueyang in 208 AD. The poet's name was familiar, and later at home I found it in the magazine *China Tourism*. There was no translator mentioned.

> *Off to the east rolls the mighty river,*
> *Washing out all traces left by great men.*
> *On the west rises an ancient rampart,*
> *Where, it is said, Zhou Yu of the Three Kingdoms*
> *Staged that famous naval battle at the Red Cliffs.*
> *Day and night terrifying white breakers*
> *Keep pounding and booming at the craggy bank,*
> *Surging and foaming the while in wild fury.*
> *How fascinating is this vast land of ours,*
> *Where countless heroes once fought and lived,*
> *Like myriads of stars shining and then vanishing.*

I found two prose poems on the Red Cliff in *Su Tung-P'o, Selections from a Sung Dynasty Poet*, translated by Burton Watson. Written in 1082 AD, they were primarily about getting drunk and meditating on the river and the moon. According to a footnote, many spots along that section of the Yangzi were called "Red Cliffs," and actually Su Dongpo did not really picnic at the exact site of the battle. Su (1037-1101) was an innovator who gave China a new style of romantic verse. He was imprisoned for 130 days and banished to "Huangzhou" in 1079 (near the battle site) as a minor government official. He had criticised the policies of prime minister Wang Anshi. Rehabilitated, he was then banished a second time to Guangdong and Hainan. In spite of his frustrations, his poetry was quite cheerful. He was also a respected essayist and calligrapher. Lin Yutang wrote a biography, *The Gay Genius* in 1947, at a time when "gay" really meant "gay." Su had three wives.

I finally found a description of the battle of the Red Cliffs in *The Romance of the Three Kingdoms*. The Han forces were led by General Cao Cao, the adopted son of an imperial eunuch. The Han empire was divided by warlords, powerful rich families, and the Yellow Turbans (184-205 AD). The Yellow Turbans were peasant revolutionaries led by religious Daoist leaders, some of them faith healers. They caused a lot of havoc. Many court eunuchs held political power too.

The famous Red Cliffs was a watershed, the beginning of the end of the Han dynasty. After its demise, from 221 AD until 265 AD, China was broken up into Three Kingdoms. One was roughly today's Sichuan Province, then known as Shu Han under Liu Bei. The other two were Wu (South China), and Wei (north China under Cao Pei, son of Cao Cao). It was a period of almost constant warfare, a time for macho heroes and military cunning.

The Three Kingdoms were glorified in song and story, passed on (with considerable trimming) by imaginative story tellers to an illiterate population, much like the Trojan War and King Arthur's defense of post-Roman Britain. Some of the heroes became gods; Guan Yu is considered the God of War and patron saint of policemen. Their personalities become prototypes. A child might be gullible like General. Cao Cao or have a temper like Zhang Fei.

The Romance of the Three Kingdom's version of the battle of the Red Cliffs might not be actual history but it is a good story. It tells of intrigue within the ranks, and a zero-visibility fog a few days before, during which Zhou Yu's men drove their boats close to Cao Cao's forces. Beating drums and shouting, they teased their enemy into shooting their arrows into bales of hay on Zhou Yu's ships. Zhou Yu collected 100,000 much-needed arrows this way.

Can't you just hear the audience laughing and clapping?

Zhou Yu also had a spy who won the confidence of Cao Cao. Inspired by the seasickness of Cao Cao's northern troops, Pang Tung suggested that Cao Cao's ships be tied together to steady the smaller ships and to allow horses and men to go from ship to ship across the river from the north shore. Cao Cao naively believed him. He is said to have had a million fighters and was overly confident of the prevailing wind. Zhou Yu's lieutenant Kung Ming (Zhu Geliang) changed the direction of the wind by using Daoist rituals. Gullible Cao Cao, expecting grain ships from Zhou Yu, allowed the enemy's ships to get too close. The ships were set afire and Cao Cao escaped with only three hundred mounted men.

Kingdoms Built on Rice

(Caroline, Entrance to Qutang Gorge, September 22, 1992) Rat-a-tat. Construction starts again at 7:00 a.m. Kailin volunteers to go for buns and *mantou*, plain steamed buns. Expired Nescafe, peanuts and apples make up the rest of breakfast. The morning's program is to see the Fengbi Wall. Meanwhile, Zhongling is to find a way to go the few kilometres to the ancient town of Baidicheng where we will meet Li.

One of the few remaining pieces of wall with an arched entrance in the Gorges area, Fengbi Wall dates from the Song Dynasty. The ancient city, called Yufu, had five gates, each with an inscription. The eastern gate was "Natural Barrier of Qutang"; the western, the "Throat of Sichuan"; the northern, "Solemn and Mighty." The smaller southern gate was, "Viewing the Waves" and the larger was, "Looking Far and Wide." This last gate was also known as Yidu Gate after some lines by Du Fu, who lived in yet another of his thatched cottages in the valley beyond: "The sun is setting behind the isolated city of Kui State. I am looking to the capital at Yidu by the southern Gate." (Yidu is the town south of Yichang, where the pottery boat thought to be connected to the legend of Lin Jun was found.)

Our guidebook says two river gates still stand: we see only one, and I like to think it is Du Fu's. In the past, access to river or town could be denied by its now vanished wooden doors. The city entrance is wider, not so grand as the arch from the river. There is no guide to point out the succeeding dynasties' contributions to the grimy masonry of the wall. But, for the wall and gate to have any meaning, they must be reproduced marking a boundary between river and town.

Fengjie is the sight of Yongan Palace, where Liu Bei is said to have installed (and later buried) the Lady Gan, his beloved concubine and the mother of his heir, Liu Chan. She was posthumously proclaimed Empress of Shu and reburied with Liu Bei (and Zhu Geliang) in a mausoleum known today as the Temple of Wuhou, in a southern suburb of Chengdu. But the buildings there now are said to date back only to the Tang Dyansty.

There is no one to show us Fengjie's sites. We are tempted by the flatbread, "flapjacks" Kailin calls them, being baked by a street peddler in a metal drum oven, but Kailin drags us along. Back at Guesthouse No. 1, a police jeep (the twin sister of the chariot we had rented in Shibaozhai) is parked in the entrance. It is for us and it will cost ¥80 to go the short distance to Baidicheng. We tell the driver there has been a misunderstanding. He insists on a ¥10 service charge, and if we don't pay soon, it will be ¥20. Kailin suspects Zhongling ordered the van although we thought he had gone off to find the *kangfulai*. IIe is twenty minutes late.

Our Party Secretary lurks nearby; Kailin cross examines the desk clerk who made the phone call. Still no Zhongling to clear things up. We must do something. It is time to save "face" for the Party Secretary and everybody else. Fearing that he will suddenly find a rule that in this county foreigners cannot take public buses, Kailin decides that ¥10 is a cheap way to dispose of the problem. When Zhongling does reappear we keep our comments to a minimum in the interest of group solidarity. Kailin and Li are getting a bit fed up with my nephew; his "love of comfort," but mostly with his insistence that he is Number Two boss. I try to point out that we are all, or none of us, bosses, except for a big emergency when I might have to decide. I do have the authority of age! If it is so important, he can be Number Five boss!

To give him credit, Zhongling has found the bus station, and we set off, single file. Our blue denim bag has long ago been proven useless, the straps torn off. We have, luckily, come upon a green rubber-backed canvas Red Army knapsack. Its straps eventually tear too, but for the meantime, it is a great improvement. Robert, Li and I have backpacks, but this army bag strikes some chord in Zhongling. No matter how heavy it gets, or awkward, he avoids carrying it when he can, and absolutely refuses to put it on his back. The bus depot is near the sports stadium. Not far, as usual, is far indeed, but this morning we are revived. True to his word, Zhongling leads us to a clean and new *kangfulai*. The trip to Baidicheng is short but these few kilometres are the most dangerous of all our bus trips. The road is indescribably rough, raw and new, and there are no safety barriers.

We follow the Meixi River, back to a point narrow enough for a bridge to be easily built, and then around to the old town of Baidi. The river mouth is the site where Zhu Geliang employed his famous Eight-Element Battle Formation in the war against Luxun. The great strategist of Shu used the vast expanse of sandy flats to his advantage to hold off a much superior force.

We have a spectacular view of boats pulling into the harbour before entering Qutang Gorge (and of water pollution from a nearby plant). The bus lets us off at a bus ticket agency. The helpful women there agree to look after our packs. We set out with cameras, lunch boxes and water bottles, for one of the main tourist sites of the Gorges. We all say no to the *huagan*: two men carring a deck chair, poles lashed on either side. In Norman's childhood important people travelled in sedan chairs in the city and *huagan* in the countryside. We will climb those stairs ourselves.

Baidicheng is a citadel, protected by water on three sides and by the Wushan Mountains behind. Formerly a walled town, it guards the break in the mountains called Kuimen Gate, the entrance to Qutang Gorge. The name comes from the Kingdom of Kui, said to have existed at the beginning of the Spring and Autumn Period (770-476 BC). The *kui*, a dragon-like creature with a single hind leg, is thought to have been the totem of a local tribe.

Baidicheng's name, White King Town, comes from the end of the Western Han Dynasty. A well in the town was said to give off white steam in the shape of a dragon. In 25 AD a local warlord, Gongsun Shu, called himself the White King and proclaimed the town, White King Town. Thus divinely protected, he set out to conquer the Kingdom of Shu, in present-day central Sichuan. Thus dragon mythology is particularly strong in this place.

Baidi today is neither a town nor a temple, but an historic site: as usual we cannot judge its age. The white dragon well is there in the garden but today, alas, there is no white steam. According to one of our guide books, in the reign of Emperor Jiajing (1522-67) of the Ming Dynasty, Gongsun Shu was demoted in favour of Liu Bei and Zhu Geliang, and the temple was rededicated to them. Mingliang Hall, the Temple of Marquis Wu and "Zhu Geliang's" Star-viewing Tower were built.

Baidi is perhaps the only city we see that was larger in antiquity than it is today. It is said to have been immense in Han times. All of southern China had eventually come under the authority of the Han emperors. Migrations of Han people to the south put pressure on the local tribes, causing what are described in the annals as "disturbances." However, many of these new arrivals had fled the Han yoke and, as soon as the empire weakened, they made alliances with the local people to carve out their own kingdoms.

By the time of the Three Kingdoms, the influx of people with knowledge of Han feudal society had provided a large enough population and sufficient economic base for south China to establish kingdoms politically independent of the north. The upstart Cao Cao of Wei had tried valiantly to gather together the remnants of the Han Empire but his defeat at the Battle of Red Cliffs in 208 left Liu Bei and Sun Quan in control of the Yangzi and of the Kingdoms of Wu and Shu, making the reunification of China impossible.

Rice cultivation was the economic base for these kingdoms. Large-scale water control projects are essential to the prosperity of the north and of the Chengdu Plain, but they are less critical in the south and along the Yangzi, where rainfall is plentiful and water storage is less important. The wetland agriculture of the Yangzi is unique in the world. Lotus, water chestnut, cattail, water-caltrop, arrowhead — all of them fleshy roots and stems, rich in starch, sugar and protein, grow naturally in marshlands or as cultivars in ponds or paddy fields, in rotation with rice. It is not that these plants are not found in other places; they are, but only in the Yangzi are they still farmed. Rice yields 96.3% digestible energy compared to 86.4% for wheat. China's prosperity in historic times has always depended on the rice crop of the Yangzi and south China.

It is a matter of national pride that rice cultivation be a Chinese "first" and archaeological evidence seems to confirm this. Historians had traditionally awarded all the laurels to the north: the Wei and Yellow River valleys were thought to be the cradles of Chinese civilization, including plant domestication. But the ancestors of domesticated rice grew only in the marshy areas of south east Asia and the Yangzi delta. It only made sense that rice domestication would have occurred where the wild plant was found.

In 1977 at Hemudu, south of Shanghai in the Hangzhou Estuary of Zhejiang Province, archaeologists unearthed a village built on piles over a peat bog. Abundant remains were found in the fourth stratum of domesticated rice (*Oryza sativa*), along with wooden and bone spades, and the bones of domesticated dog, pig and buffalo. The earliest rice-bearing level gave C^{14} dates of 7,000-6,600 BP (before the present, 1950). The nearby site of Luojiajiao yielded similar remains and a date of 7040 ± 150 BP (Zhao 1986-87:29-34).

In 1991, archaeologist Yan Wenming reported rice remains far inland, just south of the Yangzi and east of the Gorges, at Pengtoushan, in northern Hunan Province. Dated to 6000 BC, they are the oldest yet found. It is now thought that from beginnings in the middle and lower Yangzi, domesticated rice found its way to SE Asia some time after 4000 BC (Bellwood 1992:161).

Archaeologists once believed that early Neolithic people practiced slash and burn (swidden) agriculture. However the bone spades found at Hemudu are taken now to be early evidence of a kind of sedentary farming called spade tillage. There is no evidence that swidden agriculture, which has smaller yields and soon exhausts the soil, is more ancient than spade cultivation in wet areas. There is much still to be discovered about stages in the domestication of rice. One suggestion is that rice production progressed from "rainfed lowland (wetland), to rainfed upland (dryland), deepwater, irrigated, and tidal swamps." Other variables are methods of planting (broadcasting, dibbling, transplanting), water control, and the permanency of the cultivation site (Chang 1984-85:69-76). The methods and dates of rice cultivation in the earliest Neolithic sites in the Three Gorges is not yet clear. Future excavations should have much to tell us.

A dramatic group of lifelike statues in the Mingliang Hall, a wood filigree pavilion draped with banners of red cloth, tells the part of the story of Zhang Fei and his comrades that took place here. By 220 AD it seemed that the jockeying for power was over. Three states in central China, Wei, Wu and Shu, arose from the wreck of the Han empire. On December 11, after the death of his father Cao Cao, Cao Pi proclaimed himself Emperor of Wei, and on May 15, 221, Liu Bei had himself proclaimed Emperor of Shu Han at Chengdu. Cao Pi rewarded Sun Quan with the title King of Wu.

However, in 219, as the result of a temporary alliance of Sun Quan and Cao Cao, Guan Yu had been killed by Sun Quan's men, while attacking Cao Ren of Wei at Fan City north of the Yangzi. Sun Quan sent Guan Yu's head as a token to Cao Cao, but buried his corpse with all honours. Liu Bei felt bound by his oath in the Peach Garden to avenge Guan Yu's death. One by one his allies deserted him, and eventually his only access to the Kingdom of Wei was through the Gorges. Late in 222 Liu Bei's armies began taking territory and consolidating their positions along the river. Sun Quan sent offers of peace, but no apology. Liu Bei's advisors, including the faithful Zhu Geliang, counseled caution, but to no effect. Luxun, Sun Quan's general, defeated Liu Bei utterly and destroyed his army. He fled to Baidicheng.

The tableau in Mingliang Hall imagines Liu Bei on his deathbed, handing over the regency of Shu to his trusted prime minister, Zhu Geliang. His younger sons, Liu Yong and Liu Li, look on. His oldest son, Liu Chan, was left to guard Chengdu. Wisely, Zhu Geliang had refused the kingship. Although Liu Bei's dynasty was doomed, Zhu Geliang's heroic efforts kept Liu Chan on the throne, although Shu was finally absorbed by Wei in 263. Historic sources say Liu Chan was weak and worthless, but a reign of forty years in those or any times is an achievement.

Liu Bei and his friends have been treated well by posterity, perhaps much better than they deserve. Liu was a warrior hero with a reputation as an honourable man with a sense of public duty; he was also an attempted assassin. But it is for his ability to inspire loyalty in worthy men that he is remembered — and for this futile and fatal attempt to avenge the death of his friend.

The Star-viewing Tower is said to be the place where Zhu Geliang observed celestial bodies, to forecast the weather for his military operations. In folklore, Zhu Geliang was a wizard. He not only foretold the weather, he could use the secrets of Daoism to control it. His most famous exploit was to predict (or some say to divert) the winds before the Battle of the Red Cliffs, fanning the flames as they burnt Cao Cao's mighty fleet.

Our guidebook says there are 75 ancient tablets from the Sui, Yuan, Ming and Qing dynasties in the two pavilions called, according to the Chinese metaphor for these "libraries," Forests of Tablets (or Stele). Engraved on the Phoenix (or Three Kings) Stele are a pair of phoenixes, perched on peonies, underneath parasol trees: they are the three kings of nature. The Bamboo-leaf Stele integrates the art of the engraver and the art of the poet. These carefully preserved stele are carved in many scripts, representing different styles and periods. Again a fascinating array of images of Buddha and his disciples has been assembled from local temples long ago destroyed.

One room is dedicated to the universally fascinating cliff burials, and a small wooden coffin and a few human bones are on display. Murals illustrate one of the theories on how the ancient peoples transported their coffins so high up on the cliffs. Wild and primitive-looking men bearing torches cut holes in the sheer rock faces to erect temporary scaffoldings.

The guides here speak only Chinese, and they are occupied with tour groups. There is a large western-style building, perhaps the museum, perhaps housing the treasures mentioned in our guidebook. "Thousands of tools, weapons, instruments and other articles of different times," have been collected in the town: stone axes and arrowheads of the Neolithic;

bronze swords and axes of "slave society"; a series of bells from the Warring States; bronze mirrors of the Qin and Han dynasties; celadon tiger cubs from the Shu Han; and many fine gold and porcelain things from the Tang and Song. We see none of it.

Finally, it is time to meet Li at the Baidi Guesthouse. The temple grounds look out over Kuimen Gate. Kailin and I leave Robert delightedly taking pictures. The souvenir salesmen sell the local children's wooden toys: brightly-painted wooden puppets are manipulated with strings. There is a London bobby, and a Dick Wittington cat. I say they must have a Monkey King and sure enough, Kailin finds one. There are also tiny wooden tea sets.

The Baidi Guesthouse is an old-fashioned tourist establishment of gargantuan proportions; it is easy to believe foreigners stay here, but that is not much of a recommendation. Lunch is *menu fixe* and *prix fixe*: ¥60. However, we very much want to find Li and this is the only alternative to walking down that hill and then up again! We agree, but we must wait for them to cook it. Just as we are called to lunch, Li and Wang Bin of the Baidi Bureau of Relics arrive. Wang will take time from his schedule to show us a few things, we can stay at the museum guesthouse tonight and supper and breakfast will be provided. Tomorrow morning we will catch a boat to Wushan.

Wang Bin leads us the hundred metres or so to the museum's guesthouse and gives us the grand tour. Like the Shibaozhai Guesthouse, it is temple-style, but this one is a charming combination of old and new. It offers a magnificent view of the citrus groves of the Xiao (Small) Ho River Valley, and the 1800m-high mountains behind Kuimen. A cement factory belching noise and dust is the only disharmonious element. Far in the distance, we can just make out the spot where the brick memorial commemorates Du Fu's cottage.

"Baidicheng was build in AD 25, in the Western Han Dynasty," Wang says. The oldest part is on the hillside across the valley. One can, even at this distance, see where a Han street once was. The ancient city, however, is all around us and under our feet. It was destroyed in war many times and gradually abandoned. Nothing much happened in World War II. The Japanese concentrated their bombing on the local economic centre, Fengjie. We are absolutely delighted. Everything in the guesthouse is evocative of a traditional China.

On condition that we not bring our cameras, Wang will take us on a short tour of some archaeological sites. He explains that there is no excavation done here now: they have neither the money nor the expertise. He took courses in excavation methods in the museum department of South West Normal (Teachers) University, at Beipei, north of Chongqing, but he has no mandate to dig. A retired curator is a graduate of North West United University in Xian, a leading university in archaeology, but he is too old. The site is managed by the Sichuan Provincial Relics Bureau and the Beijing authorities, with Fengjie thrown in for good measure.

Wang Bin would like to collect further information about all known sites, and to then do some salvage excavation. He has heard government promises of large sums for surveys and for preservation of sites. We are confused as there seems to have been one survey after another. Wang does say that a current survey will be finished by April, 1993.

Are there many sites here that might be excavated? Many! Most are earlier than the Han. There are numerous Neolithic sites near Kuimen. The famous Daxi type-site (the site that is used to define the culture) is at the end of Qutang Gorge. Wushan County has an important Early Man site, only partially published. There are other caves as well. Two hours by bus from Fengjie is the "heavy groove," the "first gorge under heaven." And then, of course, there is the Han town of Baidi.

The tourist path winds down to the foot of Kuimen Gate. Wang Bin plucks from the clay bank beside the path what he says is a piece of Han tile. When we look closely, we also see that there are thousands of fragments. Again Han, Warring States, Xia, Shang, even Mao Dynasty, are all mixed up. (Everyone laughs at my joke.)

We leave the path, curving back around the mountainside, and into a farmer's field. Sweet potatoes are planted between orange trees. The edge of this terrace is in fact part of the old Han city wall. Centuries ago peasants filled in the gap between the wall and the mountainside, and so the ancient deposits of pottery and chipped stone tools are, in this place at least, all mixed up, not worth excavating. But there are other places that are. We go further along the old wall to a place where the hillside has recently fallen away. Here, Wang says, are undisturbed strata.

We return to the path and continue down toward the river. Wang points to our left, to the spot were the Chaotang River flows into the Yangzi. There (but now under water) are two of the famous River-locking Iron Posts, 2.13m tall. These iron pillars, affixed to the river bottom made it possible, so they say, to stretch a chain across the river and stop an invading fleet from entering Sichuan, or a merchant avoiding the tolls.

The West Wind Pavilion has a snack bar. Again we have been blessed with a warm and sunny day. Greedily we buy up the entire bottled water supply of the slow-moving girl with a small stock of drinks and snacks and sit on the stone benches in the shade. Wang Bin is eager for foreign help at fabled Baidi, but is waiting for the authorities to decide his beloved citadel's fate. The temple is safe, but it will be on an island. Baidi will lose its ancient hegemony over the entrance to Qutang Gorge and the surrounding fields will be submerged. The YVPO report speaks only of the cost of relocating subsidiary buildings in Baidi (¥100,000), and the Iron Columns, also ¥100,000.

We have taken up enough of Wang Bin's time. Robert decides to go back the way we have come to pick up the little cache of shards I have left as too heavy, beside the path, and to take more pictures. Zhongling leaves us to go back down to hire a porter and bring our luggage up to our mountain aerie. Kailin, Li and I take the direct route back to the guesthouse, up the many flights of stairs. It is hot and the steps are steep. We three women have a chance to rest and talk on the shady landings.

A shower is utmost in our minds, and a nap before supper. Our very charming but simple room has been built with bathroom attached, but the water has been turned off. At this point, I finally stop being the stupid foreigner, in this area at least. In China's small towns the best bathrooms are, not surprisingly, Chinese bathrooms, once you understand their protocol and their logic, which are ecologically sensible. This is a big step for me, who lived in China for a year, but had a Western bathroom provided by the university at great sacrifice, because the "foreigner can't bear it."

Kailin finally instructs me in the mechanics of Chinese showers and shows me the small shelf where I can put my things out of the way of the water. When you get the hang of it, it is quite pleasant. I have no companions in the shower, utter privacy. It is a great pleasure and a relief to strip down and stand under the drizzle of warm water, to wash my dirty hair and to get cool and clean all over.

We watch the lengthening shadows in the valley. Supper has been called for 6:00, but as we stroll out of our "temple" and over to the kitchen, we learn that the cook has gone to buy vegetables. There is a pavilion in a grove of trees, and nearby several men work a couple of electric saws, shaping large logs into beams for an extension of the temple. Thus we see how a temple is built, except for the power saws, in the traditional way.

A willowy young woman whom Zhongling befriended at the temple this morning joins us. She works as a guide, having been assigned to work here upon leaving junior high school. She would like to learn English because she is sure that the dam will bring many English-speaking tourists to their museum. Zhongling has a chance to come to the rescue of a pretty girl. He brings out his English textbooks, carried all this way, but as yet unopened. She might listen to the lessons on BBC or VOA, or even try to get herself sent to SISU. Thank you, she says, clearly, and goes down the hill to her parent's home.

Supper is finally ready. A table has been set in the same room as the simple kitchen. An 8-year-old is doing her homework, close beside us, writing characters, looking up characters in three books, one at a time, including a dictionary. So much work to learn to write! A much smaller girl, perhaps five, looks on. These children are poorly clothed and badly need a bath before bed. The little one hangs around her sister; neither show any interest in us. Or perhaps they are very shy, quite different from our friends' small emperors and empresses.

Supper is simple indeed. Braised cabbage, pork and green peppers yet again, and a good and tender chicken. We went slowly back to the guest-house. The stars were beautiful. We had not seen any in the cloudy skies of Chongqing, but here the air is clear. The valley is quiet and we are almost alone in our temple. Should we let down the mosquito nets? It is a lot of trouble for someone to put them up again. Instead, we keep the windows closed. This is not a good choice: it is stuffy in our little room!

Great Poets All Visited the Gorges

(Kailin, Qutang Gorge, September 23, 1992) When first our ship sailed into the Gorges area, Kuimen Gate loomed in front of us, straight as a ramrod, as beautiful as a picture. Every foreign and Chinese visitor gasped in admiration. One of the visitors from Guangdong shouted: "Three Gorges, how I love you!" All through the Gorges, we felt as if we might as well be sailing through an ancient painting of water and mountains. The peaks tower sky high, the winding river cuts deeply into the valley. Upwards to the sky, downwards to the torrent, the peaks reach to the heavens while the ship runs in a crevasse. Looking forward from the prow, we wonder is there an escape through the mountains? And yet, when we turn the corner, we find ourselves under another sky. Many poets and painters have found in the Three Gorges the enduring themes of countless works of art. A good poem must be magnificently conceived.

Robert displayed his talent in drawing at Qutang Gorge. Later I came to know that he was trained in drawing from childhood and that this skill played a role in getting his books published. He asked Huang Zhongling to translate the poem *On Leaving Baidicheng* for him and asked me to write Kuimen on his drawing in the five styles of Chinese calligraphy: *Lishu, Zhuanshu, Xingshu* (running hand); *Kaishu* (regular script); and *Caoshu* (cursive script).

After we discussed it a while, suddenly he stopped talking. To my surprise, I found that he blushed. He was weeping with tears in his eyes, like a small child. This is the first time in my life that I saw a grown man crying. I asked him why. He rubbed his eyes with his handkerchief, for quite a while. We kept silent for several seconds. Then he said, "Kuimen is the most beautiful thing that I have ever seen in my life." Never before have I felt so strongly the beauty of the mountains and rivers of our lovely country. People make light of this charm, although they travel a thousand *li* to visit it. It is this charm that made the person from Guangdong shout like a crazy person. It is this charm that so excites poets including Robert.

Scholars have divided the poems of the Three Gorges into three kinds: descriptions of nature with political meaning, human history and poems in praise of particular places.

Poems that describe the natural wonders of the Gorges to express political ideas or serve political purposes: *Gao Tang Fu* and *Fairy Fu* (*Fu* means a graceful essay) were written by Song Yu to express his frustrations because, politically, he could not achieve his ambition. In the same vein, *Qu Yuan Lie Zhuan* (the Record of Qu Yuan's lifetime) in *Shi Ji,* by Si Maqian, conveyed Qu Yuan's feelings of love and hate and praised his noble aspirations, to which he remained faithful and unyielding. At the same time, the poem exposed the corruption, and muddleheadedness of the aristocratic clique of the Chu State, which eventually harmed the country and ruined its people.

The Entrance-of-Three-Travellers Cave by Bai Juyi, describes Sanyou Cave near Yichang. "This scenery is wonderful," wrote Bai Juyi, but he used the "poetic soul" of the mountains and rivers to express his despair and the bitterness of his lot. Even on sighting Qutang Gorge, he moodily wrote, "If you want to know how great my worry is, well it is higher than Yanyudui (Goosetail Rock near Baidicheng)." Ou Yangxiou was demoted and exiled there. The poets Lu You, Yu Lu, Lu Wen and Yu Wen all borrowed this method of describing and using historical sites to express their own anger and indignation to the rulers of the Tang and Song Dynasties. In fact the poets had five methods of dispelling the "tragic consciousness of existence": plunge one's self into nature, seek a fairy's blessing, drink alcohol, dream a beautiful dream, or play chess.

The second type are poems that describe the history of human activities and of civilizations, or disseminate scientific knowledge, history, geography, politics, military lore, local customs and specialities, hydrology and shipping, and the weather. A good example is *The Annotated Classics on Water* in which Li Daoyuan recorded all of the mountains and passes the Yangzi River passes through. It is a magnificent source for the study of history, geography, and water conservancy in ancient times. It is a veritable encyclopaedia of great scientific and literary value. Ou Yangxioiu's record of Xiling Gorge in ancient times has also yielded a treasure of historic materials.

The third style are poems that sing the praises of scenic spots and historic sites along the Three Gorges, using them to express patriotic feelings. Li Bai's quatrain, *On Leaving Baidicheng*, which is also known as *Going to Jianglin*, is a depiction of the marvellous natural phenomena of the Gorges. On his way to exile to Yelang, Li Bai was stopped in the Gorges and told he had been pardoned. Excited, 27 or 28 years old and at the height of his powers and not yet bearing the scars of failure, he returned to Jianglin, writing his famous poem, which perfectly expresses youth's vigour and optimism. The night, when it was getting dark and we took the boat across the river to Shibaozhai, there were no stars in the sky. When we got off, only the lamplight from the top of the boat lit our way to the road. It was so quiet that I could imagine the poetic situation in which Li Bai wrote his poem, talking about his light boat slipping past ten thousand ranges.

The grandness of Qutang Gorge cannot fail to strike a chord in one's heart. It brings home to me an important aspect of Chinese poetics. Chinese landscape poetry often strives to present nature in its pure and original forms, and the viewer is thus brought into closer contact with nature itself, and with the state of mind that inspired the poet.

In *Dawn over the Mountains* there is a famous line, "The still sky... the sound of falling leaves." I like this evocation of tranquillity most. If I could have stayed here, away from the noise of the city, I would never have suffered from insomnia. The day we stayed in the hostel of the Baidicheng

Administration for Protection of Historical Relics was a fine day. The hostel has a commanding position, and I feasted my eyes on the magnificent horizon. Facing the beautiful rivers and mountains, breathing in the tranquillity and the fresh air, I wanted to sing out loudly "I love China," but my voice was hoarse from a sore throat. That dark night, the stars in the sky and the lights like stars in the valley of Baidi Town formed another wonderful picture. Look at the surface of the river! I felt the "poetic situation" of *Night at the Pavilion Ge Ye*, by Du Fu. "In the Three Gorges, the reflections of the stars and of the Milky Way, quiver on the water."

It was quiet and peaceful in the hostel, and outside tall pines added a solemn touch to the ancient structure. I could imagine that when the great poet Du Fu made his abode in Kuizhou for two years, he lodged many a time in the vicinity of Baidi City, in these extremely beauteous surroundings, where legends and scenic spots inspire nostalgia and prompt creative thought. He expressed this in four hundred or so verses, threaded through with awakened recollections and yearning meditations. *Ascending, A Night at the Pavilion* and *The Fall of Golden Leaves* are the most celebrated. And the peak of poetic perfection is considered to be the line, "Raindrops fall as if from heaven, streams of the Yangzi murmur through all eternities."

One of the most famous modern poems about the Yangzi is, of course, Mao Zedong's verse, *Swimming*, written in 1956.

> *I have just drunk the Water of Changsha*
> And come to eat the fish of Wuchang
> Now I am swimming across the great Yangzi
> Looking after to the open sky of Chu
> Let the wind blow and waves beat
> better far than idly strolling in a courtyard
> Today I am at ease
> It was by a stream that the Master said —
> "Thus do things flow away!"
> Sails move with the wind
> Tortoise and smoke are still
> Great plans are afoot:
> A bridge will fly to span the North and South
> Turning a deep chasm into a through fare
> Walls of stone will stand upstream to the west
> To hold back Wushan's clouds and rain
> Till a smooth lake rises in the narrow gorges
> The mountain goddess if she is still there
> Will marvel at a world so changed

There is an old saying, "There must be a poem while travelling to Wushan." It implies Wushan is so beautiful that everyone might produce the inspiration to write a poem. These legends and poems are also the inspiration for painting, music, drama, and dance.

The brush and ink, which painters use to create another world, are characteristic of Chinese painting. Trees, stones, currents, streams, rivers, and gardens are brought vividly to life before us. A large mountain and water mural entitled *The Yangzi River Eastward* was designed and executed by a team of art workers led by Chang Ding, on a wall at Xizhimen Subway Stop in Beijing. The frescoes depict a bird's-eye view of the Yangzi from Chongqing, past Wanxian, through the Three Gorges to Wuhan, proudly depicting the large water conservancy project, the Gezhouba Dam. It blends the appeal of an ink line drawing with the dynamism and motion of the "splash-ink" technique of painting.

Plastic arts or the art of modeling are also expressed at Baidi Citadel. There are statues of Liu Bei and Zhu Geliang that are lifesize, full colour depictions of these historical figures. The Chinese poems in the temple galleries are unique. These protective screens can be put in the middle of a hall or restaurant (we saw them at Yunyang, Zhongzhou and elsewhere) to partition a place or give shelter from the breeze. The screens can be folded or arranged with furniture, and they make a limited space rich and colourful. Li Bai wrote in a poem, "When visiting Yuan Dan Qou, sit down by the screen of Wu Shao."

Here, at Baidi temple, there is a painting of Wushan that looks very like the Wushan we visited the next day. I see once again the twelve mountains of Wushan on the horizon. The screen increases my enjoyment in reading the poems. If you ask, "Which mountain or mount did the poets most praise?" perhaps I would answer Fairy Mount. According to the official calculation, up to the time of the Tang Dynasty, more than a thousand poems had been written about it. Song Yu, Kiu Hui, Li Bai, Du Fu, Bai Juyi, Liu Yuxi, Mong Xiau, Li Huo, Li Shanying, Wei Zhang, Shi Maguang, Wang Anshi, Lu You, Wand Wei, Fan Chengda and Su Shi all wrote poems in praise of Fairy Mount.

Perhaps Liu Yuxi's poem best expresses why so many poets chose to praise it. "Whatever its height, so long as celestials and immortals exist, Fairy Mount will be famous." Fairy Mount, or sometimes Beauty Mount, rises a not very lofty 912m above sea level on the north side of the Yangzi at Wuxia. Atop it is a pillar of stone, ten metres high, thought to enclose the soul of the Fairy Yao Ji, who helped Yu Da to harness the Three Gorges. When our boat passed by, the loudspeaker's voice drew our attention to it. "This is Fairy Mount! Usually she is unhappy to see foreigners. But today she is moved by your sincere yearning and pure friendship. She welcomes you. Today is a lucky day."

Standing high and looking into the distance at the lovely scenery of the Xiling Gorge, I remembered my visit last year to Yichang. I listened from my balcony there and could hear the graceful and heart-stirring sound of classical music. It was perhaps because I cherished an ardent love for music that I rushed into the concert hall. Oh! it was so sweet that I was absorbed in it. A group of young musicians performed with ancient musical instruments from archaeological excavations with such exotic names as *Bianzhong, Bianqin, Fangxian, Yun,* zithers and inverted bell. The actors and actresses wore ancient costumes. When they were moving the strings of the zither, striking the chime of the bronze bells, the music intoxicated me. Even at present, whenever I recall that beautiful music, I regret not to have listened to it a bit longer, that day. Even my mother, who was with me, said, "It is no wonder that such great poets as Bai Juyi, Yuan Zhen and Bai Xingjian all visited the Gorges. "

Into The Gorges

(Robert, Baidicheng to Wushan, September 23, 1992) We rise from our comfortable mosquito-netted beds in the splendid Baidi Guest House. The sun is already shining over the orange grove terraces that surround our mountaintop. Its light plays off the upturned eaves of the building that seems to me to be a summer palace fit for the emperors of old. The great chasm of the Qutang Gorge that opens before us is still deep in shadow. We breakfast on steamed noodles while sitting on the steps outside the kitchen. The workers we sit beside will soon take up their tasks of guiding the boatloads of tourists and maintaining the site. This is yet another example of how our "connections" are allowing us to experience things that ordinary tourists would not see. The bill for the best accommodations we will have on the whole trip, along with the breakfast and last night's dinner, is embarrassingly low, less than ¥60 ($12US).

We are all in high spirits as we wend our way down the steep stairway toward the boats far below. For the second day the sky is blue and the day promises to be clear and bright. We are almost alone going down but two boatloads of Chinese tourists are clambering their way up. This time we have no prepurchased tickets and will just have to take our chances on getting passage.

The two boats that are moored on the mud flats at the bottom of Baidicheng's precipitous cliff are different from any that we have yet been on. They are smaller steamers, with three instead of four decks, squarish and with rounded bows. They look more like classic North American river boats than the other Yangzi ferries. The negotiations, as always, are protracted and way beyond Caroline's and my understanding. These boats are already filled with people on packaged tours and casual, port-to-port fares are obviously not the rule. The results, however, are successful. For a relatively small amount, about ¥35 ($7US), all five of us will be conveyed through the first of the mighty gorges to the next stop at Wushan. There, we hope to travel up the Daning River to see what are known as the Three Small Gorges.

Although we have no cabin space on the boat, Zhongling manages to get us keys to the lounge on the top deck. There we pile our gear and make ourselves at home. Li pulls out a reclining deck chair and proceeds to curl up for a nap. Not me. The thought of finally having reached the gorges themselves has me out on the open deck, cameras at the ready.

The boat swings out into the turbulent current of the river's main channel. If we had been here perhaps twenty years ago we would have seen a huge rock projecting up out of the river and dividing it into two passages. *Yanyudui* or Goosetail Rock had been one of the most dangerous features of the Yangzi since time immemorial. It was most treacherous during the summer flood season when it was awash. Like so many other places of importance along the great river, the Goosetail Rock was surrounded by stories. One of these legends holds that the rock stood above the Dragon Castle. One day, long ago, *Long Wang*, the Dragon King, decided to leave his palace under the rock and tour his underwater domain. This he did against the advice of his ministers. He transformed himself into a fish and was so distracted by the wonderful things he was seeing that he became careless and was caught in a fisherman's net. Soon he was being sold in the market, perhaps in one of the same streets we visited yesterday in nearby Fengjie.

The legend goes on to say that before he knew it the woman who had bought him had him split and frying in her wok. I could believe this because we had seen fish taken live from buckets and cooked with a great concern for freshness. Nevertheless the story says that the Dragon King, when he did disengage himself, went directly to the palace of *Ya Hwang*, the chief ruler of the gods, and complained about the way he had been treated. *Ya Hwang* told the Dragon King that since he had left his proper place and turned himself into a fish he had only himself to blame for his misfortune.

While a story like this may have its underlying Confucian ethic about staying in our assigned stations in life, the same standards don't seem to be applied by the Chinese to natural features. The Goosetail Rock was a hazard and it was decided that it had to go. In 1958 it was blown up, and now it is only a memory.

The regular passengers line the railings ready to snap pictures of themselves and friends, arm-in-arm before the great sights of the Three Gorges. This scene is almost ruined by the nature of the boat's design. The short funnels, one on either side of the boat's open promenade deck, constantly spew acrid fuel oil fumes. The funnels are not tall enough to carry the smoke clear of the deck. The result is that, in the swirling air currents of this place which is known literally as the "Wind Box Gorge," the choking wisps of black soot encircle first one group of picture takers and then another. But not even this annoyance can detract from the excitement.

As we approach the gorge I look in vain along the riverbank below Baidicheng for any sign of the Iron Posts that are said to have once held chains that stretched across the river. These items were listed in the Canada Yangzi Joint Venture Cultural Heritage survey as dating from the Song Dynasty, 960 to 1279 AD. They are scheduled to be moved. Other readings of mine, however, have indicated that, far from being ancient, they may date only from the reign of Emperor Daoguang in the Qing Dynasty, 1821 to 1851 AD. What strikes me is not how old they might be or whether they ought to be moved — that can be accomplished easily enough — but how these iron posts could possibly have been very useful in holding a chain to control river traffic when they are under water most of the time?

The morning sun is just clearing the peaks that frame the Kuimen Gate. The pink stone wall on the north side burns with light. The sheer south face, still in shadow, is black by contrast. The rays glint off the boat's bow wave, and illuminate the eddies and whirlpools that announce the river's violent current. Each change in the boat's course reveals another vista of vertical rock. We shoot along on the racing current. Here the whole mighty flow of the Yangzi, collected from the melting ice of the Himalayas, the rivers that drain the vast Sichuan plain and hundreds of named and unnamed tributaries, are channelled into a rock-walled passage hardly wide enough for two ships to pass. This is truly a place of poetry and legend. What had been awe inspiring the day before, when seen from the heights of Baidicheng is exhilarating when experienced from the surface of the river. After a flurry of shooting I let the cameras hang from my neck and allow my senses to fix the place in memory. My eyes scan the rugged cliffs, the wind pushes against my face; through my feet I can feel the surge of the boat through the water; I can hear, if not understand the excited chatter of the holiday-goers and I can smell the efforts of the machine that carries us over the current.

In the midst of the wonder of this general spectacle there is one particular detail that manages to stand out. Cut in the solid stone on the north wall of the gorge is the most striking example we have yet seen of what is called a "trackers gallery." This particular one may date only from the last century but there are many remnants of ones that are much older. Trackers were the men who used to be employed in huge numbers to tow boats upriver against the current. I know from books that the practice lasted until well into this century, and I will find out later in the day that it survives in some forms even now. So remarkable was the sight of hundreds of small, incredibly tough, wiry men hauling a 120 tonne junk through these rapids with nothing but their cooperative and aggregate use of individual strength and stout bamboo hawsers, that virtually every European traveller who ever passed this way left a colourful account. Now, on the main river anyway, there is nothing left to testify to that extraordinary process that went on for hundreds of years except this scar across the rock face. Just a hand-hewn pathway, perhaps a metre and a half deep and barely tall enough for a man to stand up. In this season I guess that it stands about ten metres above the waterline. Also, by my calculation, the rising waters of the planned Three Gorges reservoir will put this trackers gallery well below the new water level.

One of the three proposed reasons for building the Three Gorges Dam is to improve navigation on the Yangzi. Because these stretches of fast-flowing and turbulent water will be transformed into nothing more than narrow spots on a slack-water lake, the need for boats to fight their way upstream, the attendant dangers of manoeuvring and the potential for collision will be eliminated. At least, that is the argument. I have seen some reports that propose measures to improve traffic control and navigation aids. In addition, the same proposals suggest the increased use of winching stations whereby the age-old system of human tracker power is replaced with a modern adaptation — cables and mechanical devices. It is felt by some observers of the Yangzi that such improvements could accomplish just as much increase in river traffic tonnage and safety at a far lower cost, both in capital investment and cost to the environment. I think of the computer-operated traffic control systems on the St. Lawrence Seaway in Canada. There, operators, like air-traffic controllers, use radar, radio signals and TV monitors to keep track of every vessel in the waterway. The Chinese can put rockets in space, and the navigation system on the Yangzi is already a vast improvement over what it was in the past. I can't help feeling that if improving navigation were the real goal, they could do it without building this dam.

The use of winches to haul boats upstream has a slightly different meaning for me. As with other things, winching represents an indigenous solution to a problem. It would be the logical modern adaptation of a local tradition to suit available technology. It would be a home-grown rather than imported answer to a challenge. My own feeling is that indigenous and local knowledge approaches are generally the better and more sustainable ones for developing countries.

But observations such as mine are not what the boat's PA system is encouraging among the other passengers. Now that we are in the Gorges proper there seems to be an endless stream of poetic and mythological references to the landscape. Kailin finds me and pursues her appointed task of translating the loudspeaker comments. She also points out the features on the shore. Soon we are passing what she describes as Meng Liang's Ladder. Sure enough, on the south face of the gorge I can see the line of holes carved in the rock and disappearing upward as far as one can see. One of the legends from the Song Dynasty (960-1279 AD), is that General Meng Liang and his men were trapped here in the Wind Box Gorge. Patiently they cut holes in the rock and inserted timbers horizontally as rungs by which they ascended the cliff and surprised their enemies. It is said that cutting each hole cost ten lives. One hopes that the industrial safety record for the Three Gorges Dam project is better. A variation on the story tells of a bet between

Meng Liang and another general about whether Meng could get over the cliff by morning. In that story, Meng is tricked by a monk who makes the sound of a cock crowing and convinces the general that he has failed to reach the heights in time.

One of the things that makes these legends interesting is that all the other traditions concerning Meng Liang place his activities far to the north of Sichuan. Did he really make the remarkable march to the Three Gorges, or have beautiful places simply demanded stories of famous people to populate them in their literary dimension? I came to realize that every kilometre of the Three Gorges is as densely populated by legend as China is by people. "There," Kailin points, "there is the Upside-Down Monk, the one General Meng kicked off the top of the mountain when he found out he had been tricked." There is a rock that I guess is shaped a little like a crumpled-up monk after a serious accident. I know right now I am never going to be able to keep all the stories and anthropomorphic boulders straight. Later I read that the holes of Meng Liang's Ladder only went about half-way up the rock face. They were more likely used in past times to support ropes for people climbing to collect valuable herbs from the cliff. But the whole wonderful effect of the place is not diminished in the least by checking the facts.

The Qutang Gorge is the shortest of the Three Gorges but, perhaps because it is the first one I see, for me it is the most spectacular. After a few kilometres our boat emerges from the eastern end of the steep-sided defile. The mountains are still high but their peaks rise further back from the river. Villages, farmsteads and small factories again appear along the somewhat less rugged shoreline. The day remains sunny and clear and is beginning to turn hot. I break out my Tilley hat and squint into the distance looking for the next town.

Wushan soon comes up on the left side of the river. It is quite different in its appearance from the places we have visited further west. Instead of rising directly out of the river as does Fuling, Fengdu, Wanxian, Yunyang and Fengjie, this town is set back from the main channel of the river. Between town and riverbank are sloping red clay flats that are exposed at this time of year and in places are a hundred or more metres across. These flats front on the Yangzi and wrap themselves around the town, extending up the tributary Daning River as well. What is most distinctive as well as quite disconcerting about this place, is that where the edge of the town does rise sharply from the river level the community is busily dumping its garbage. This is taking place not in any localized area, but is general all along the edge of the town. My guess is that the process has been under way for some time. The fires that are burning in places along the dump, however, are probably of more recent origin. We are to have the opportunity of a closer look at this phenomenon later in this eventful day.

Our boat goes past Wushan almost entirely before turning into the mouth of the Daning River. The line between the waters of the murky

Yangzi and the relatively clear tributary is sharp. The boat noses into the muddy bank among several others of various sizes, and is made fast. Several small launches appear almost immediately and pull up alongside. These are steel-hulled vessels about 15m in length with a kind of open wheelhouse projecting up at the stern. They seem to have retractable tops over double rows of seats. They are obviously the famed Small Three Gorges tour craft. The trouble is, we are told, there are just enough spaces for the tour-package passengers. There is certainly not enough room for the five of us. We can hire another boat, we are told, and have it all to ourselves. That will cost ¥600 ($120US). No thanks. This part of the Yangzi tourist infrastructure is clearly well organized and efficient. We are asked to wait aboard until the regular passengers have all disembarked. Then we jump ashore onto the slippery red clay and begin to consider our next move. Will we be able to get to the Small Gorges after all?

9

WUSHAN

Wushan Man And Daxi People

Caroline, Wushan County, September 23, 1992) I awake with a start. Li has turned on the lights. She is sick and needs Kailin to help her find the bathroom in the dark. When she opens the door, mosquitoes come in. I pull the quilt over my head but it is no use. Dawn is just breaking. I load the video camera to film the sun rising over the Xiaoho Valley but when I get to the balcony, the sun is already too high. Li returns, looking better, but out of breath. Our plans have changed. The boat will leave at 9:00 a.m. and so we must get up immediately.

The cook will make us noodles. His wife and daughters are eating already and the temple peddlers are picking up their umbrellas and boxes of goods. We have the run of his kitchen. Quickly we assemble eggs, chives, salt and spices from his small pots and watch as he fans a package of wheat noodles into his steaming wok. As they cook he scoops off the froth. Do we like them hard or soft? Hard.

A Chinese kitchen has low counters and simple sinks made of concrete and ceramic tiles; it takes room to chop the ingredients and lay out the dishes and their sauces. Cut vegetables and meat are temporarily stored on bamboo shelves, on bamboo platters. Most woks in Sichuan are heated with natural gas, although some are coal-fired. The cook puts noodles in my bowl. I add a spoonful of soya sauce, salt, garlic, chives, and there we are, a wonderful breakfast. For the final touch of authenticity, Robert squats on his haunches to eat, but that does not last long. Then it is "gear up" and down those long, long steps to the ferry dock.

The tourists we will join are still up looking at the temple, their boat having been run aground on the flat rocks to let them off. The local people are setting up their stands to sell them breakfast on their way back. Zhongling goes to negotiate our tickets and Robert and I sit down on the rocks, enjoying the early morning sunlight and the hustle and bustle. Several goats clamber about on the rocks. Goats are uncommon in this part of China: are they here to clean up the tourist scraps? Far across the river is a stone tower. An ancient fortification? No, a brick kiln. Someone is blasting in the river. I videotape the goats, the boom, and the slow rise and fall of the plume of gravel and dust and water.

Zhongling returns. This boat has been booked by vacationers from Liaoning Province and they have taken all but the steerage class tickets but, for ¥20, we can camp in the ballroom. This ferry is specially designed for local Chinese tour groups: not a luxury liner, but not a regular ferry by any measure. None of the Liaoningers seem to notice us or mind that we have suddenly appeared among them. And so, we will finally see the first two gorges.

We pile our gear into the ballroom and soon Kailin is happily resting on a bamboo mat on a foldout cot. The rest of us grab a deck chair and set it up out on the observation deck to see the sights in style. Time for sun screen. A blast of the whistle and we are off. Smoke and cinders blow back into our eyes. The smokestack is too low. No matter where we move the deck chairs, the cinders follow.

Never mind. Off left and right railings are many of the most fabled sites of the Gorges: the trackers' road, Phoenix Spring, rock formations, a plank road, temples, caves. Fishermen are sheltering in the coves made by rock cliffs. Our three friends photograph each other in front of the most famous sites. This is Zhongling's first trip through the Gorges and he is thrilled.

The valley widens out, the hillsides are terraced once again. We are about to sail by two of the most archaeologically important sites in the Gorges. Wushan County's *Longgupo* "Dragon Bone Hill" cave site is about fifteen kilometres south of the Yangzi. It was excavated in 1985-8 by a team from the Chinese Academy of Sciences and the Chongqing Natural History Museum, led by the famous physical anthropologist Huang Wanpo. In the karst (limestone) cave were found together fossils of the extinct *Gigantopithecus blacki* and parts of a mandible (jaw bone), a premolar, a molar and an incisor thought to have come from two *Homo erectus* individuals, along with a retouched flake tool and a hammerstone. Both tools show evidence of having been used.

Longgupo is interesting for three reasons: its early proposed date; the discovery of *H. erectus* and *Gigantopithecus* fossils together in a stratified cave context (the only similar find was in southwest Hubei at Jianshi); and, finally, the discovery of stone tools with both.

H. erectus used simple tools and eventually learned to control fire. Its brain size approaches ours. The first hominid to have left Africa, it has been found in the Middle East, Indonesia (Java Man) and, most notably, at Zhoukoudian near Beijing (Peking Man). In fact Peking Man, found in 1929 in excavations led by a Canadian, Dr. Davidson Black, are the most extensive *H. erectus* remains we have yet found. Tragically they were lost during the Second World War and are known today only from the excellent casts made by physical anthropologist Franz Weidenreich who, not surprisingly, is the father of the multi-regional theory of human evolution.

Studies of the bone and teeth found at Longgupo show that "Wushan Man" may have been an intermediate type between Java Man and Peking Man. The date of the site is still controversial: various new methods yield conflicting dates between 2.04 million years ago (mya) by paleomagnetism (measuring reversals of the earth's magnetic field), and 2.39 mya by amino acid racimization, a test done on the bones themselves. An analysis of the animal bones associated with these finds (the biostratigraphy) confirms that the date is early Lower Pleistocene (about 2 mya). Wushan Man's nearest contender for the oldest human remains found in China is Yuanmou Man, dated by paleomagnetism to 1.7 mya, found in 1965 in Yunnan.

Gigantopithecus blacki was first named by Ralph von Koenigswald in 1935, after huge lower molar teeth he found in a Hong Kong drugstore. More were found soon after in a stratified context in a cave in Guanxi. Now there are thousands. Based on the size of its teeth, it was first thought that this primate was about three metres tall. (One guidebook mentions legends of Yeti in the Wushan Mountains and so, if one is a Big Foot believer, it neatly ties together.) However, it has since been shown that our primate four-chambered heart cannot sustain a body that large, and from other species it is inferred that *Gigantopithecus* was probably about the size of a female gorilla — with very large thick-enameled teeth. The teeth chewed a rough vegetarian diet that must have been largely composed of twigs, leaves, seeds and nuts.

Gigantopithecus is a hominoid, but is it a hominid? It was previously thought (even by Franz Weidenreich) to have been directly on the evolutionary path that led to the genus *Homo*, or at least to *Australopithecus*. One reason for its demotion is simply its survival as a contemporary of *H. erectus*. The teeth found in China and elsewhere are much too late (Early or mid-Pleistocene) for their owner to have been on our evolutionary branch. The consensus today is that it belonged to the Family Pongidae, but was an evolutionary dead end.

When and where each subgroup of our primate ancestors left the main trunk of the tree of primate evolution is still a matter of fierce debate. (The evolutionary tree, as Stephen Jay Gould points out, looks more like a bush than a tree.) The famous missing link, the ancestor of us all, is a shadowy construct defined, it seems to me, as "not anything we have ever found a piece of."

Since Raymond Dart discovered his Taung Child in South Africa in 1924 and established the genus *Australopithecus* as ancestral to *Homo*; Don Johanson, Lucy, *Australopithecus afarensis*; Mary and Louis Leakey, *Homo habilis*; and Richard Leakey, Turkana Boy *Homo erectus*, the African hominids have held centre stage. So far the only accepted very early specimens of our human lineage have been found in eastern and southern Africa.

But the Chinese, spurred by national pride (and little else it sometimes seems), have held that our human ancestors will eventually be found, if not in China, at least in Asia. Their most famous paleoarchaeologist Wu Rukang, in a lecture in 1981, argued that: "Fossil sites of *Ramapithecus* have so far been mostly found in Asia and the earliest sites are those in Turkey and India. If *Ramapithecus* is a hominid, then it is most probable that the birth place of man is in Asia."

Key to the Chinese argument is showing the development from the *Ramapithecus*, or *Sivapithecus* as it is more properly known, through *Australopithecus*, to *H. erectus*. In the published version of his lecture, there are pictures of "four teeth suggestive of *Australopithecus*, found in western Hubei in 1970" but with the qualifier "this is not certain" (Wu 1982:14). Three were found with *Gigantopithecus* remains. The fourth, the most controversial, turned up in a Badong drugstore. It has some similarities to australopithecine teeth and caused a great stir of excitement. However in his book in 1985, Wu says that statistical analysis places the tooth beyond the accepted range for *Australopithecus* but within that established for *H. erectus* (Wu 1985:74).

Where does our Wushan Man fit into all this? The article in *Asian Perspectives* comments: "Many important questions concerning the course of human evolution in South China and the extinction of *Gigantopithecus* in the Middle Pleistocene might be answered through continued excavations at Longgupo and other localities in Wushan County, Sichuan, and adjacent territory in Hubei Province. If the reported chronometric dates for Longgupo are correct, not only would these hominid remains be the earliest known outside of Africa, they might well provide the evidence necessary to understand the degree to which human evolution in East Asia must be considered a phenomenon distinct from that of regions to the west" (Olsen and Miller-Antonio 1992).

Wushan Man must have been part of the earliest migrations of *H. erectus*, out of Africa, through SE Asia and north into China, much older than Peking Man, who is, after all, less than 500,000 years old. In fact, if his date is correct (it is very controversial), Wushan Man is among the earliest *H. erectus* found anywhere! As for *Sivapithecus*, it is most probably the ancestor of the Orang although there is still disagreement. Unquestionably, there were many primates, now extinct, occupying this part of the world, between Turkey and the South China Sea, 16 million years ago.

Although assigned to our genus, *H. erectus* is still quite different from living human beings. When and where, then, did anatomically modern *Homo sapiens* develop? From the widespread populations of *H. erectus* or in one place? Until recently, the one place theory held sway, and that place was thought to be Africa. Dubbed the "Noah's Arc" or the "Garden of

Eden" hypothesis, the theory says that we are a new species that appeared less than 200,000 years ago. We spread rapidly, replacing indigenous populations without interbreeding with them. Then we settled down and developed (very quickly, it seems) the different characteristics of the modern races.

This theory was bolstered in the 1980s by data based on the analysis of mitochondrial DNA from human placentas. Geneticists from the University of California studied the mtDNA that is only passed along the female line, arguing that we are all descended from one woman, a "Mitochondrial Eve," who lived in Africa.

A bitter war of words broke out between the geneticists and the anthropologists, and for a while the geneticists seemed to have it all their away. But the bones experts, led by Milford Wolpoff, a pugnacious paleoanthropologist from the University of Chicago, chipped away at their methodology and at last count had sent the geneticists back to their labs with their tails between their legs to check, among other things, their "molecular clocks." The arguments are immensely complex and the technology of the geneticists is so new that they will undoubtedly be back for another round.

For the meantime, the Chinese (and Wolpoff) and the theory of multiregional evolution have retaken centre stage. Their task now is to "fill in the fossil record." They believe that the Mongoloid race developed "in place" from Peking Man and his relatives. Their theory is based on a list of skeletal and dental peculiarities (notably "shovel-shaped" incisors) found not only in *H. erectus* but in modern Mongoloid populations. It makes more sense, as Wolpoff is fond of pointing out, that the local population developed indigenously than to assume that the invading *H. sapiens*, who most certainly were quite different, developed the same local peculiarities as the ancient skeletons, and in only a hundred thousand years.

Nowhere in the world, in the transition period from *H. erectus* to modern humans, do we find neat statistical or typological patterns. Human evolution is a mosaic, and it is "best to view all Middle Pleistocene hominids [730,000-128,000 years ago] in a broad perspective as an essential part of one evolving lineage in direct ancestry to modern humans" (Wood 1992).

So far there is a relative lack of data in the years 75-25,000 years ago. Cranial parts and teeth dating from 25,000 years ago have been found in south and central China and in the Sichuan Plain; the Yangzi and its tributaries then and now, are the only way through the mountains.

China only recently assumed its present geography. Fifteen thousand or so years ago, you could have walked to Taiwan, or Japan or the Philippines. Just 7,000 years ago, this huge continental shelf was finally submerged by the melting waters of the polar ice cap. Only ten or twelve thousand years ago did the Tibetan Plateau achieve its current elevation.

European archaeologists divide the cultures of the Middle and Late Pleistocene into the (beautiful) Acheulian Hand Axe Culture of Africa, the

Middle East and Europe and the not-so-beautiful Chopper-Chopping Tool Culture of Asia. French archaeologists working in Vietnam in the 1920s identified a Mesolithic culture they named Hoabinhian after a site on the Red River. It is now recognized over the whole of SE Asia, including Japan. It stretches back at least 40,000 years and it is a chopping-tool culture, arguing again for local continuity. About 15,000 years ago plant remains become more prominent and some say that gardening had begun. About 12,000 years ago, pottery begins — in Japan. About 10,000 years ago agriculture begins and with it, the Neolithic.

The earliest sites found are in caves (settlements on what is now the continental shelf would today be under water). In southern China they have been pretty well mined by peasant apothecaries looking for "dragon bones" and few have their stratigraphy intact enough to give reliable dates. It has been argued, but not yet really documented, that the earliest cultures of the Gorges and the Sichuan Plain were Hoabinhian-like.

Future prospects for paleoanthropology in the Three Gorges Area are exciting. It is not just that valuable caves and other living sites will be submerged (certainly they should be explored). Much larger areas of rough and mountainous terrain will be accessible for the first time. If only the archaeologists can get there before the druggists looking for "dragon bones!"

Li points to a terrace at a river mouth on the south bank. Daxi, a painted pottery culture, roughly contemporary with Yangshao Culture, was first found here in 1958. Daxi people occupied this and other sites from 4500-3100 BC, a considerable period of time. Daxi is the "type site," the site used as the basis for defining the culture. Similar assemblages of artifacts found here and elsewhere had a definite enough geographical and temporal distribution to justify the decision to declare that a new previously-unknown culture had been found. Archaeologists have a natural human proclivity to "my sititis": to think the culture of "my" site is unique (and thus merits a separate name); and that "my" culture influenced other cultures. It is the old story of ego and of the "splitters" against the "lumpers," although distinguishing preliterate cultures often seems a matter of opinion.

Neolithic people lived on broad terraces above the normal flood level of the tributary rivers, where they would have easy access to the water and to fish in the river and agricultural land. Only a few sites, like Yangjiawuang and Baimiao, are on the hillsides further back. There are three main Daxi settlement areas: at Daxi; at Xianxi, east of Zigui, and at Miaohe, including Zhongbaodao Island. A thick layer of sandy silt at both Zhongbaodao and Daxi is evidence of a huge flood of the Yangzi at both sites about 2500 BC. It allows archaeologists to date both sites by comparing their stratigraphy. So far, we knew that Daxi Culture existed in the eastern tip of Sichuan (Wushan County), in northern Hunan Province (Tangjiagang Type) and in western Hubei (Guanmiaoshan Type).

Although fish was a mainstay of their diet, the people at Daxi grew rice. ("Western" scholars tend to think the Chinese pay too little attention to botanical remains.) The relationships between Daxi Culture and the rice-growing cultures of Hemudu and Pengtoushan downstream are not yet clearly understood. Burials were found from both early and late phases in a community cemetery, in narrow graves the size of the corpse. The earliest people buried ground stone and bone tools with their dead; the later ones, pottery and necklaces and bracelets of jade, stone, and bone.

Archaeologist Li Wenjie, however, distinguishes the Daxi way of life from that of the contemporary (and better understood) "Yangshao" cultures in the Central Plains: "The Daxi people were primarily rice growers in contrast to the dryland farmers of the Yangshao culture who primarily grew millet. Furthermore the Daxi people widely used native wood and bamboo for constructing different parts of their buildings. In pottery making, a fine bamboo double-edged bi-comb was used for making the comb-tooth designs. And it appears from the woven designs on the bottom of pottery vessels and the bamboo tube-shaped *ping*-jar that bamboo mats and tubes were used in everyday life" (1979 2:161-64).

Excavators assign the earliest periods at Daxi to the matrilineal stage of human development and the later ones to the initial patrilineal stage. Chinese archaeology has been shaped by the philosophy of Marxism-Leninism, which in turn is based on Hegel, Lewis Henry Morgan, and Edward Tylor's notions of the development of human society, popular in the 19th century. Human development is divided into periods of primitive communism (or communalism), slave society, feudalism and then capitalism. Chinese scholars, with varying success, have sought to apply this paradigm from Greece and Rome to China. As well, they support the theory that early societies, clan society in China, were matrilineal. Patriarchy took over when advancing technology permitted private property to triumph over communalism.

The earliest periods of human development had a matrilineal lineage system. The clan lived communally, technology was primitive and productivity was low. There was little division of labour, and no private property. Women had the main responsibility for agriculture under the system of spade tillage. Men and women were buried singly, with their clan, in densely-packed community cemeteries. Grave goods were few and large inequalities of wealth are not evident from the burials.

The middle period was a time of transition. Production methods improved and craft specialization developed. As surpluses increased, households of couples came to be the main economic unit, accumulating private property. Husbands and wives were buried together within their clan plots. Inequalities in quantity and quality of grave goods are increasingly found, indicating that differences in social status were becoming significant.

In the later period, private ownership was well developed and disparities in wealth and social class are clear from grave goods and from grave sizes. Exquisite handicrafts in jade, ivory and stone, as well as wheel-made pottery, suggest craft specialization and well organized trade. Agriculture was controlled by men, and with it the economy. Men, clearly the heads of households now, are sometimes buried with several women, their wives and concubines. The transition to patriarchy is well under way and the primacy clan system is giving way to the broader community. Villages increase in size and ditches are supplanted with walls, built for military purposes.

Sounds logical? Proving it from the "archaeological record" is something yet again. Did this change come about gradually, or through "revolution?" Nowhere is the evidence neat or complete. Was there ever a matriarchal society? A slave society? When did the clan system disappear? Are the marks on Neolithic pottery the precursors of writing? Do they indicate a very early beginning of "civilization" — complex societies — in China?

The watermarks on the banks here say 57m, then 32m: they do not make much sense. Our little group must talk. It is suggested that we hitch a ride with these tourists up the Small Three Gorges of the Daning River, stay overnight on their boat at Wushan, and ride down with them to Yichang. We are almost out of time. I am reluctant to cut short our trip, but I am worried about Norman. It seems probable that there are no berths available on the ship we are on and we do not know but that these people might want to use their ballroom for dancing! We are even more unclear how we will get back from Yichang, and we have not been able to cash any of our traveler's cheques so we are running low on cash.

Until I actually help buy the tickets, I cannot believe that the return trip by boat is three nights and two days. We have no good choice. Zhongling would like to go on to Wuhan, stay with his uncle, and fly back to Chongqing. But it is a day's train ride or 24 hours on the boat from Yichang to Wuhan and we have no tickets for either. And then, there are only a couple of flights a week from Wuhan to Chongqing. There is a train from Yichang to Chongqing but it requires a major change, and as we cannot buy a through ticket we might be stuck for a day.

We are approaching Wushan County town. An immense expanse of mud flats stretches below the first road of the little town. When we arrived here in 1988, it had just rained. The steps were covered with mud and small rivulets of water ran everywhere, including down the stairs. At the top we boarded buses and were taken through the town (a trip still vivid in my memory) to a launching place on the Daning River, just in front of a huge stone bridge.

Our boat, however, steams past the steps and beaches itself on a sand bar. The small tourist boats that converge on us look bigger than I remember them and so my confusion begins. Zhongling goes down to negotiate joining our tourists on one of these boats. Communication is complicated by the fact that the captain locks the doors to the lower deck so the boats can be boarded with a semblance of order (Chinese hate to line up). Time passes. More time passes. Finally it seems the last tourist has gone. Word finally comes from Zhongling via Kailin: Every seat is full, prebooked.

I remembered talking to a former tracker, via Norman's Sichuan dialect, who was now peavey-man on one of these "willow leaf" boats. He proudly quoted statistics on how much their business would expand in the coming few years. He looked deceptively young, but in his lifetime he had gone from beast of burden to business man, and the future he saw was bright!

Robert and I have been happily filming the rowed and power craft leaving and entering the Daning. Now, if we do not do something soon (it is 11:00am and getting hot), it will be too late to go up the Small Three Gorges and we will miss one of the most exciting experiences on the trip. We have two choices: hire one of these boats; or, walk over the immense mud flats (a kilometer, with all our luggage) and up the stairs (beside the burning garbage heap), find a bus or other conveyance, go to the bridge and there look for a boat. Kailin goes to find the captain or tell Zhongling to get a porter, or something. The captain is still too busy to deal with the problems of his (probably) illegal passengers and all of the porters have left.

If we must walk, we might as well get going. But first we must get out onto those mud flats. We yell at Zhongling to get someone to open the door. Zhongling has befriended an old woman who says she will find us a porter. She picks up a half dozen of our bags and sets off through the treacherous canyons of mud eroded by the runoff. About a hundred metres on, she waves down an old man rowing a sampan. We are absolutely delighted to get in. He eventually drops us at the stairs. Our benefactress grabs the bags again and starts looking for a local boat.

We will need a boat with a good motor. She waves down a cargo boat loaded with pottery and two baskets of chicks. We agree on ¥100 to take us to the top of the gorges. This exciting possibility has made Li and I forget all about our time constraints. Li says that we can get another boat at Dachong, further up the river, for another small town where she knows there are interesting things. We can stay the night and return tomorrow. Yours truly, who wants just once to go farther than foreigners usually go, is thrilled.

The Small Three Gorges

(Robert, Daning River, September 23, 1992) So here we stand on these vast red clay river flats. The beautiful clear waters of the Daning River issue from a V-shaped cleft in the mountains before us. The town dump of Wushan smoulders behind us. Half a dozen tourist ships, one brand new with the word "Police" written in English along its side, line the riverbank. But we are on our own.

"When I was here before," Caroline says, "I think we got the excursion boats up there by that bridge."

In the distance we can see a rainbow-arched bridge high above the Daning. It seems to be hanging almost in defiance of gravity from the sides of two mountains. I remember once again studying the Tang poets in their all too inadequate translations. Li Bai's verse *The Road to Shu Is Steep,* comes to mind:

> *How dangerous, how high!*
> *It would be easier to climb to heaven, than walk the Sichuan Road.*
> *. . . sky ladders and hanging bridges*
> *Above, high beacons of rock turn back the chariots of the sun.*
> *Below, whirling eddies that meet clashing torrents and turn away. . .*

The others think we might be able to get passage on a local boat if we can get across the mud flats and closer to the town where the smaller boats seemed to be clustered. I have no opinion but I am not too anxious to see Dante's Dump first hand. Anything is preferable to walking up to the town proper. It is already so hot just from the sun that we are drenched in sweat. A middle-aged woman in typical peasant blue shirt and pants approaches us and asks if she can be of service. We need a porter and a boat to take us up to the town.

"She will carry our things," Zhongling translates, "and find us a boat. . . ¥5 ($1US)."

Caroline protests as the woman, who is older than her, is loaded up with the better part of our gear.

"Think about it," I say to Caroline, "you wouldn't want somebody to say you can't publish books and run a business because you're a woman. This woman's probably worked like this all her life." I'm not totally comfortable with this position but it seems to me that the woman deserves some sort of respect. It is a question, however, that didn't seem to concern our Chinese friends in the least.

As it is we do not go far. When we get to the water's edge the woman hails one of the passing boats. It is an open flat-bottomed affair about six metres long and a metre and a half wide in the middle. Standing in the stern is an ancient brother of the Yangzi. He is dressed in blue trousers and a white smock, his skin tanned like leather. In the river breeze wisps of grey hair blow across the crown of his bald head and trail from the corners of his enigmatic smile. For the first time I am not only going to see a traditional Yangzi boat up close, but to ride in one. This moment of our trip passes quickly enough but the image of the old man, like the Charonesque character from one of a thousand tales about the mystical crossing of rivers, sticks firmly in my mind.

With his right hand he propels the boat by the use of one long oar. It crosses in front of him and works against a post on the port, or left side of the craft. His left hand rests on what at first appears to me to be a most unwieldy tiller or steering device. This rudder consists of a couple of log-sized timbers as big around as a person's upper leg. It projects far beyond the stern. But the most unlikely feature of all is the boulder lashed to the end of the oar nearest our boatman. Then I realize the rock is a counter-weight and that the seemingly awkward steering mechanism is actually perfectly balanced. Furthermore, its considerable weight gives it the stabil-ity required to keep the boat on a true course. The old man's oar strokes are slow, rhythmic and powerful enough to move us at a leisurely but ade-quate pace. Steering requires only slight pressure on the tree trunk-and-stone tiller. There is a spare oar and another rowing post near the bow of the boat. I have to check a desire to pick it up and try my hand.

We are put ashore close to Wushan's small boat port. I busily photo-graph several of the craft, heavily laden with stones and being rowed by three men each. We are soon able to attract the attention of a boat that is setting off up the Daning River. It steers toward the shore and a shouted conversation ensues. Yes, they will take us up the river. . .as far as we want to go. . . ¥100 ($20US). This is a much better deal than the alternative, or so it seems at the time. The boat noses into the shore and we all jump aboard.

"Down under!" Zhongling translates the excited shouts of the boat's crew. "The foreigners have to stay down here." Caroline and I tumble down the steps from the open front deck of the boat into the covered central sec-tion. There we find ourselves among a jumble of assorted cargo: plastic pots, cardboard boxes and crates of baby chickens. There are also a few other Chinese passengers, clearly amused by our unceremonious arrival.

The boat is about 15m long. The cargo hold is perhaps two and a half metres wide. The boat is flat-bottomed and constructed of wooden planks fastened together with what appeared to me to be wrought iron staples or cleats. There is no keel. A large and noisy gasoline engine mounted inboard near the stern powers the craft. The roof over the cargo hold is sheet metal supported by wooden and steel posts. Flimsy curtains flutter over the open-ing between the top of the planking and the roof. Zhongling appears down the steps.

"It's OK now, you can come up. We've passed the police boat." Appar-ently these guys are not supposed to take foreign tourists up the river.

The front deck of our boat is big enough for all of us to sit comfortably. There are also two of the crew. Before long we will find out what they are here for. At the very front the hull of the boat comes up into an elegant and, I think, entirely Chinese shape. In the centre of this squared-off flare there is a stout iron nail about as big around as a man's thumb. The use of that item too will become clearer as the day progresses. Hooked over the bow of the boat is a four-pronged wrought iron anchor. Lying on the plank seat beside me and projecting back up over the roof on the cargo hold are two bamboo poles at least five metres long. Fastened to their ends are what look like iron spearheads. All these pieces of hardware are the kinds of things that are probably made in the workshops and blacksmith forges we have seen in several of the riverside towns we've passed.

Soon we are passing under the hanging bridge. The road it carries over this first gorge is at least a hundred metres above the valley. Then, as we round a bend, the river suddenly narrows. Upstream, in front of us the surface of the water boils. What we are approaching is not so much a set of rapids as a small waterfall, perhaps a metre in height.

The boat's engine roars and fumes as the helmsman heads us straight for the smoothest patch of water available. Two of the crew who have been chatting with Zhongling jump up and grab the five metre bamboo pikes. For all its noise the boat's engine can barely hold us stationary against the current once we have reached the sharpest part of the waterfall. The two crewmen in front of me then thrust their poles into the water and begin to push. They lean and strain and bend their whole weight against the bamboo poles. When one comes to the end of his pike, the other holds while the first quickly pulls up and probes the bottom for another foothold. Slowly the boat inches forward until, finally, the top of the rapids is reached and the boat again surges forward under its mechanical engine alone. The boatmen sit down, hardly breathing hard.

Along the stone walls of the gorge there is yet more evidence of the fabled plank roads: a double series of holes about 20cm square and perhaps two metres apart. In former times timbers would have been placed in the top row of these holes to project horizontally from the cliff face. These were probably then braced with another timber angling down to the holes in the row underneath. Between these brackets the surface, or planks, of the road would have been laid. These holes are said to date back as far as the Song Dynasty in the 12th century.

Kailin has been talking to one of the passengers. She slides over to my seat to explain the local tradition about the spot we are approaching. "Here are the gold and the silver piles," she says. "In the long history there are many landlords and rich persons who have been turned over here, so that under the river there are piles of treasure such as gold and silver."

As we round a bend in the Daning and emerge from the first of the Small Gorges we come upon a scene that represents another of those folds in time or vistas from another age. We have read about the trackers pulling junks up through the Yangzi narrows. We have seen old photographs of the trackers galleries cut in the walls along the river. Now, beside us as we reach the second rapids, we see an open wooden boat, its ends swept up in graceful curves, and from its bow long ropes stretch forward. At the ends of these lines men strain against the power of the current. They are tracking the boat upstream as their ancestors have done for over two thousand years.

The definitive writing on the water craft of central China, in English at any rate, is a monumental book entitled *Junks and Sampans of the Yangzi River*, by G. R. C. Worcester. Worcester was an Englishman who spent most of his working career as a customs inspector in China, and shared the fate of many Chinese officials when he was interned in a Japanese prison camp during the Second World War. Writing in the 1960s he lamented the passing of many traditional styles of boat and speculated that before long most of the old boats would disappear. Worcester died before his massive work, with its beautiful diagrammes and colourful stories, was published in the 1970s. What would hopefully have given some comfort to him is that many of the traditional-style boats have survived the intervening twenty years.

Studying my photographs and descriptions of these boats after returning to Canada I realize that the stone-carrying boats at Wushan were what Worcester categorized as Kulinto coal junks (*Gu lintuo*). They were developed to transport coal down to Wushan from a mining area further up the Yangzi. The first small boat we rode in was an undersized version of the coal junk, but with many of the same characteristics. Part of Worcester's description mentions "...the very distinctive stern sweep," or oar, in which, "the balance is nicely, if crudely, adjusted by means of stones lashed on the upper surface of the loom end." While I don't see any of these boats loaded with coal, they are being used along the Daning to carry stones and gravel dredged from the river bottom.

The most common style of larger wooden boat that I see on the Yangzi River itself is something between what Worcester called an *Yichang cargo junk* and the *new all-purpose junk*. It is perhaps not surprising, in a state that has experienced a lot of central planning and enforced conformity, that the more modern versions of the ancient river craft are a rather plain reflection of their forebears. Worcester called the *Yichang* boat "uninspiring" and said it had little originality. The *all-purpose junk* is said to be the first new design in centuries, and was introduced in the 1930s under government auspices to provide a standard and uniform carrier. Gone from these boats are the elegant lines of the old junks. They measure anywhere from 12 to 20 metres in length, about four metres in the beam and are perhaps two metres in depth. There is almost always a deckhouse of some sort, which extends about a third of the way forward from the stern.

On later reflection I realize that the boat that took us up the Daning was a motorized version of the traditional fan-tailed junk (*shenbozhi*). The overall dimension and design are the same as the Daning boat, which is said to have been first built by an almost god-like monk in distant antiquity. Although its design features are surrounded by what Worcester called a "superstitious aura" they can now be analyzed and understood in terms of their pure functionality. The only thing that is missing from the modern version is the gondola-like stern from which the older boat took its name.

Originally this feature may have served to keep water from spilling in when the boat was shooting rapids, and also may have been useful as a hand-hold for crewmen who often had to jump into the water and manoeuvre the boat around river obstacles.

That is exactly what the men at the second rapids of the Daning are doing with the upturned stern of their boat as they track it upstream. It is an example of the third style of boat from Worcester's catalogue that I am able to identify. The *sichuan* or snake boat was probably so named because it is relatively narrow, tapered and turned up sharply at both ends. As we travel further up the river the steep walls of the first gorge give way to a broader valley, where fields and garden plots stretch down in places to the water's edge. Close to small villages some of the graceful snake boats appear to be moored using "stick-in-the-mud" anchors, a pole fitted into a hole in the hull of a boat where it sweeps up at either the bow or stern, thrust securely into a muddy spot in the bottom of the river.

The only other boat I see on the main river that comes from Worcester's typology is a version of the *Wuxi fishing sampan*. There are literally hundreds of these tiny craft patiently working the waters of the Yangzi for the much-prized catfish. Some of the fishing sampans I see were fitted with long booms from which huge dip nets were suspended. These nets are hauled by a man working a treadmill in the centre of the boat.

Before we reach the third set of rapids we pass one of the regular tour boats. It has pulled up to the shore and is letting off its passengers. They can be seen climbing a path up the bank in front of them.

"What's going on?" I ask Zhongling. He translates my question. "The tour boats aren't allowed to carry people over the next rapids," comes the answer. "Too dangerous. They walk over that hill and get picked up on the far side."

I can hardly wait to see what's next. We come around a bend and there, sure enough, is another stretch of foaming white water. This time it is about twice the drop of the previous two. The engine roars and before I know it we are into the boiling froth. Slowly we make our way forward. Then, suddenly, one of the boatmen in front of me loses his hold on the bamboo pike. He lunges for it but in an instant it is swept away by the current. The boat's forward motion stops. We begin to lose ground, to turn sideways to the current, to heel over. The man at the helm has to make a quick decision. He does. We turn sharply and then run back down the rapids. The shouting and pointing that follows is even more excited and wild than we have become used to among the ever-excitable Sichuanese. The pole bobs up first here and then there among the waves. It is swept ahead of us toward the far shore. We give chase. Then in the distance we can see a boy who seems to be fishing from the rocks by the river. He sees the bamboo pole coming his way. Into the water he plunges and soon has the fugitive pike in his grasp. He swims to shore and pulls it and himself out. By this time we are close enough to shout. The boy holds up his hand with five fingers outstretched.

"He wants ¥5 ($1US)," Zhongling reports.

The men on the boat shout more excitedly than before. "No," they will not pay to get back their own pole. But we are running downstream and are unable to pull ashore. We shoot past the boy and as we do so he begins to run up the slope behind him, carrying the pole. Slowly the boat turns and makes its way back to the spot where the boy had stood. The four-pronged anchor goes ashore and the young man who has lost the pole dashes up the steep bank, his bare feet oblivious to the tangles of underbrush and sharp stones. But the boy is gone. We wait, and the excitement calms.

"This kind of thing never used to happen in China," Kailin says almost apologetically. "Now everyone wants to make money. It's all people think about."

The young man who has given chase returns empty-handed. We pull away from the shore and he slumps dejectedly in the bottom of the boat. This is obviously a serious situation for him. There is nothing I can say or do. I wonder how we will now get over the rapids. When we reach white water again both boatmen lean their strength against the remaining pike. Still, we seem to be going backwards. Several times the boat almost turns sideways to the current. I give some thought as to which way I will wade ashore if it comes to that and to how much of my camera gear I might rescue. I rearrange my bags on the deck beside me. It keeps me busy. Eventually, however, we gain the upper part of the stream. The water flattens and once again we are on our way, still making better time than the regular tourists whom we pass reembarking on their legal craft.

The second of the Small Three Gorges is longer than the first, and the adjacent peaks even higher. For the first time since I have been in China I get the feeling of really being in wild country, uninhabited by humans. I begin to see colourful birds. I have seen very few birds at all in the settled valleys. On flat slabs of rock near the river, grain has been sprinkled to attract monkeys for the gratification of the tourists. Around the next bend the tactic is working. Swinging down from the trees I see the only wild animals I am to encounter in China. As quickly as they appear, they are gone. Caroline misses them completely when she tries to get the video camera organized. All she gets is pictures of trees.

While we are in the second Small Gorge I am treated to another sight. Much is written about the famed hanging coffins in Yangzi guidebooks, but in reality they are difficult to find. The local people we are with attempt to point one out. And then I see it. A couple of hundred metres above us and about three-quarters of the way to the top of the mountain, clearly visible in a narrow cleft of rock, an unmistakable shape.

When we leave the second of the small gorges the valley again broadens and signs of intense cultivation reappear. Before long a large village comes up on the north bank of the river. This place has a kind of stone esplanade along its river frontage and stairs up to its whitewashed buildings. Along the esplanade several of the tour boats are tied up. This is obviously the terminus of the Small Three Gorges trip which is, in fact, only two gorges worth. We expect to go on though. Along the steps there are many vendors selling fresh fruit which looks quite good. With their customary excitement over such things Kailin and Caroline go to inspect the offerings.

I have seen a boatyard just beyond the stone walkway where a couple of new craft are under construction, so I am off to investigate. As it turns out, these are not traditional wooden boats, but rather welded metal versions with essentially the same design as the older ones. In many places along the main river and other tributaries I see boats being constructed and repaired, generally in makeshift sites such as this one, rather than formal yards. The semi-tropical climate of Sichuan makes permanent buildings unnecessary, and huge fluctuations in the rivers' water levels make construction on the banks impractical. I take a close-up photograph of one of wooden boats here on the Daning, which is my favourite picture from the whole trip.

I return to the esplanade to find that all hell has broken loose. The Sichuan tempers are as hot as the sun. It seems that the crew of our boat now doesn't want to take us any further up the river. They have suddenly remembered that the area beyond this village is off-limits to foreigners. They have quietly told Zhongling they don't want to take us any further but Caroline has just found out. And now they want the full ¥100 ($20US) fare they had originally asked for. Zhongling is in a lather because he doesn't like losing arguments. Caroline is disappointed since she'd had her heart set on going further up the Daning than she, or perhaps any foreigner recently, had gone. Kailin is embarrassed by the attempted price-gouging of some of her fellow countrymen.

I soon realize that our boatmen are not going to oblige us, regardless of the outcome of this discussion. This village is quite pleasant, but there is clearly no accommodation here and I doubt that the situation will improve further up the river, although Caroline says there is supposed to be a larger town ahead. The tour boats are beginning to leave to go back down the river. It looks like there might be some room aboard the last one. I take out ¥80 ($16US) and give it to the leader of the boatmen. I am glad I don't understand what he says. I am also happy that it is broad daylight and that there are many witnesses around. The last tour boat agrees to take us all back for ¥50 ($10US) so we are not too badly off.

When we are aboard I am again given the prime photography spot in the bow of the boat. This time the craft is one of the steel-hulled variety but, like the wooden ones, it is fitted with the flared bow with the iron post in the centre. We pull away from the stone wall and turn into the middle of the river. The trip down will be much faster since we will be running with the current. Now the boatmen pull out a long oar that has been stowed along the side of the boat's retractable roof. It is equipped with a bracket that fits around the iron nail in the bow. The art of going downstream is now to be demonstrated. In the calmer stretches of water the bow oar is suspended above the surface. When we come to bends in the river or rapids the two boatmen in front of me skilfully drop the oar into the blue-green water and pull or push it to bring us sharply onto a new course. Without these sharp turns the boat would shoot forward and crash into the rocks.

As we plunge and wind our way back through the spectacular peaks and cliffs of the Daning Gorges I think about the effects that the imminent flooding will have on this valley. As with the other scenic panoramas along the Yangzi, the flooding will diminish, but not destroy, the value of the place. The best of the arable land will be drowned. Some of the orchards that supply the excellent pears Kailin has bought will be destroyed. The riverbed stone and gravel will become inaccessible. Potential archaeological sites that are exposed as the river changes its course will be lost. But I think the greatest effect of the Three Gorges Dam reservoir will be on the small boats that have survived the centuries, especially in these secluded valleys.

It is my opinion, and a reflection of my own interests, that the wooden boats of the Yangzi are one of the principal material expressions of the region's culture. People travel on them, work on them and even live on them. There is a continuous tradition of both boat building and boat handling that stretches back to the place where history mingles with legend. The boats are crudely built by Western standards but they are both beautiful in appearance and functional in design. Long experimentation has seen the marriage of the most suitable and (up until now) available raw materials, with the best size and cut of boat for the particular local conditions. Hardwoods such as *qin kang* (oak) and *nan mu* (a kind of laurel) are combined in clever ways with softer species such as *shan mu* (fir) and *song mu* (pine). The planks of these boats are fastened together with a minimal number of wrought iron cleats and spikes and the spaces between them filled with bamboo shavings. That has allowed the creation of craft that are strong enough to stand the rigours of work in rough currents, but which have until now been economical to build. The snake boat we saw being manhandled through the rapids near the beginning of our day's adventure, for example, is just the right length and weight for the four-man crew to push upstream. The boat we have ridden in is just large enough for the two boatmen to pole up over the waterfalls.

Yet now there are at least three major threats to the continuation of this rich water-borne tradition. Only two of these have to do directly with the building of the Three Gorges Dam. The first factor is the expected increase in the size of ships on the Yangzi. With the opening of the huge new locks that are planned for the dam and the deepening and widening of the river into a lake, it will be possible for bulk-carrying barges, and even self-propelled ships a hundred or more metres in length, to navigate as far upriver as Chongqing. That will probably make most of these small carriers uneconomical.

That is exactly what has happened in other parts of the world. I have spent most of my life close to the Great Lakes in central North America. I know their history well and have written extensively about ships. There were thousands of smaller vessels, both wood and metal-hulled, sail and steam-powered, carrying a wide variety of cargoes well into this century. For decades after the introduction of quite large ships in the mid-19th century, the smaller boats competed. They disappeared gradually until the time the St. Lawrence Seaway opened the Great Lakes to international shipping in the 1950s. Now the same tonnage as a hundred years ago is carried in a few dozen giant ships and not a single wooden-hulled sailing vessel remains of the vast fleet that once spread their canvas over the Lakes.

A similar situation exists along the Atlantic coast of France. There, in dozens of small ports over hundreds of years, a wonderful variety of specialized, locally designed and constructed vessels proliferated. Each harbour, with its particular wind conditions, industries and fishing needs developed its own unique craft, much the same as each tributary of the Yangzi has done. In 1986 a movement began in France to build replicas of the old-style boats because virtually none of the originals remained.

It might be argued that the unique small boats of Sichuan and Hubei will be able to move further up the tributaries and continue their useful lives there. The rising lake, however, will drown most of the river-bottom resources that the boats are used to exploit as well as the farm land that produces the goods and sustains the people they carry. From what I see, as well, few of the tributary rivers are very navigable above the areas that will be flooded. Without a viable economic reason to build and operate the boats there will be little incentive to keep them, and even a China with new economic strength will not, for some time, produce many people able to afford boats for recreational purposes.

Hand in hand with this shift in economic viability will go a decrease in the interest to learn the crafts of boat building and handling. I wonder if the particular experiences we have had on the Daning River are harbingers of the future. The person who most impressed me with his skill and grace was the old man who rowed us along the shore at Wushan, while it was a young man, perhaps from lack of training or experience, who lost his pole over the side of his boat. Will he have the time, opportunity and inclination to grow into a true river man?

The greatest threat to the continuation of small wooden boat building on the Yangzi, however, has nothing to do with the Three Gorges Dam. Wooden boat construction may cease because there may be no wood left.

What strikes me in particular about the countryside of Sichuan is the absence of trees. Many of the hillsides and mountain slopes are completely bare. In the valleys there are often only a few small trees clustered around the farm buildings and sparse copses at the very tops of hills. Only in the wildest stretches of the Daning, where the monkeys appeared, were the trees thick.

The facts about deforestation in China are staggering. In the last generation the country's timber resources have been ravaged. A number of factors have contributed to the situation. In the 1950s the economic development initiative known as the *Great Leap Forward* stressed the need for iron making. All over China trees were felled to make charcoal for backyard furnaces. The chaos of the Cultural Revolution led to a widespread breakdown in both authority and the distribution of goods. Desperately poor peasants in the countryside had little alternative to cutting trees when no other fuel was available. More recently a policy of expanding grain production led to further clearing of land. Sichuan, which once possessed some of China's most extensive forest reserves, has lost almost a third of its trees since the 1950s.

Although I see hundreds of wooden boats in use, and even some under construction, I see very few standing trees of a size that could provide the timbers used in boat building. There are a few vessels loaded with timber, making their way downriver. I assume that the mountains of western Sichuan are still able to provide some trees sufficiently large that the three to five metre planks used in boat building can be cut from them. But one wonders how long that can continue.

There is an irony to the possible demise of the Yangzi wooden boats, as with many aspects of what passes for progress in the less developed world. Wooden boats may be replaced by steel ones in part because of a lack of wood. Except one of the main components in steel production is coal. In China, one of the principal consumers of wood is the coal mining industry. Almost all China's coal is mined underground, as is evident from the many pits we see along the Yangzi. Each 1,000 tonnes of coal extracted requires 13 to 25 cubic metres of timber for pit props. I wonder if there is much gain in using the wood to hold up the ceiling of a coal shaft so a boat can be made of steel, rather than just building the boat from wood in the first place?

Perhaps I am just being hopelessly romantic about beautiful old vessels. For the men pulling the snake boat upstream against the rapids there is little romance. For the old man patiently rowing his coal junk there is probably a hope of a better life. And if there is a better future for the people who toil on the boats I see then they have every right to seek it. What is lost in terms of their ancient culture might be a price they are willing to pay.

But the culture that these boats represent doesn't necessarily need to be lost forever in the push for progress. A useful role for foreigners who appreciate what the traditional craft mean might be to show their interest by assisting the local people to preserve some of their past. The preservation of artifacts such as boats and the maintenance of construction and handling skills cannot be a priority of people involved in the fundamental struggle for economic survival. It may be legitimate for people from presently more prosperous societies to aid in this aspect of development. Such aid might take the form of collecting and storing some outstanding examples of wooden boats. It might take the shape of helping local people in a town such as Wushan to set up a museum or interpretive centre for boats. In the future any such efforts may well have some payback in terms of tourism, but initially it should be done simply because it's a good idea.

Boat Burials and Crag Coffins

(Kailin, Daning River, September 23, 1992) As we went through the Three Gorges, we could see the rural people's approach to the proper disposal and veneration of their dead. We could see people burning incense, paper houses, paper made to resemble money. These were burned to pay tribute to the deceased on the burying grounds, on the sides or on the tops of hills and mountains. The rural cemeteries are modest in size, very often they are only a pile of earth.

There are a wide variety of burial styles in the Three Gorges including: Earth burial, practiced by more than thirty nationalities; cremation, which prevails among the Han; water burials where bodies of the dead are given to the river; wild burials which expose the body to the elements, similar to celestial burials; and finally, crag burials in which coffins are placed on high crags.

Crag burial is one of the characteristics of southern Chinese minority cultures, and an important part of the ancient culture of the Three Gorges area between Fengjie and Yichang. There are three theories as to how these peoples got the coffins up so high on sheer cliff faces.

One involved cutting holes in the cliff face and building a scaffolding on which to rest the coffin. Another involved climbing to a natural crevice or cave and leaving the coffin there. The body is then said to be serenely buried on the crag. Still another would involve the digging of a small cave, putting half the coffin into it, leaving the other half exposed. This is called crag cave burial. In a crag burial, the coffin is described in many ways, such as a boat, a case, a box, a bellows, or a cupboard or cabinet. The Bo people often stood their coffins on end, so burial was vertical. Legend described these burial places as the places there fairies were reborn.

When we first arrived in the Gorges the boat announcer's voice called me back. "Now, look at the right bank of the Qutang Gorge. There is a cave in the crag, about one hundred metres above the surface of the river and seventy metres from the top of the crag, where legend says that Mu Guiying (a heroine of the Northern Song Dynasty) hid her books on the art of war, along with her suits of armour, in the so-called Armour Cave. Also on the cliff, in 1958, were found three coffins, some bones and funeral objects, including a Ba-style bronze sword." The voice continued: "Now look at the left bank of Qutang Gorge. There is a wood coffin like a bellows in a crevice on the crag, which is thus called The Bellows Gorge or The Wind Box Gorge. This is one of the most marvellous sights of the Gorges, but today most of the coffins have decayed and fallen into the river. In the winter of 1971, however, the cultural centre of Fengjie found two more crag coffins here."

When our ship came to Binshubaojian Gorge in Zigui County, (*binshu* means book on the art of war, *baojian* means double-edged sword) we can again see traces of crag coffins. According to legend, this was the place where Zhu Geliang (the great statesman and strategist of the Three King-doms who became a symbol of resourcefulness and wisdom in Chinese folk-lore) hid his books on the art of war and his sword. In fact it is an ancient crag burial place. Many of these legends involve the hiding of books. Erding Crag, opposite Wuqi county town, was also described as a cupboard for hiding books. Again, however, it was a crag burial place. There are more burials in the caves and crags of the Yangzi's tributaries, including the Dan-ing.

Why did people of the Yue and Pu nationalities practice crag burial? (*Yue* is a general term for Bai Yue people living by the ocean). *Pu* is a general term for Bai Pu people living in the southwest mountain area of China. Are their burial practices related to their worship of a mountain god? So far, no one can answer. We are still finding vertically placed coffins. Whether this simply suited the terrain, or they just fell down, or whether it had a special cultural connotation, is not known. There are many hypothe-ses, but no proof and no agreement.

In a recent report, "Boat Burial, a Conundrum," it is argued that the "boat" was the thing left behind by the deceased. Others argue that the "boat" was on the crag, the deep stream beneath it was the Milky Way, and the fairies rowed boats down and along the Milky Way on their way to earth, to our land of China. Later the fairies abandoned the boats and flew back into the sky. So the "boat coffin" was the "fairy's boat."

Why did the ancients use boats as coffins? Boats were important transportation tools, and so the boat became an important symbol. People would need a boat to survive in another world. Some scholars have further suggested that the Bai Yue might have thought their souls would be taken back to their ancestral lands in these boats, there to find true peace.

Others have inferred that putting the boat coffin into a cave on a crag was expressing respect for the dead. The deceased would have been a per-son of high regard in the tribe, perhaps the chief. High on the cliff, the dead chief would have held a commanding position still, safe from enemies and from wild animals. As the Bai Yue worshipped their ancestors and their dead heroes, the safety of their dead was related to the safety of their tribe, of the living. And so the tremendous effort of raising these coffins so high can be understood.

The Bai Yue had no crane or pulley. How could they lift a thousand kilograms or more, including a boat, coffin and body into a cave or onto a wooden platform? Some have argued that in ancient times the flow of water was greater, especially in flood times, and the water level was higher, closer to the cave entrances. The boat might almost have been paddled in. Later the waters subsided and at low water the boat coffin was safe. These burials may be indications of changes in water levels.

A second opinion is that the Bai Yue set bamboo scaffolding against and into the cliff face. In the middle and lower reaches of the Yangzi, bamboo trees grew to great size, and the Bai Yue had remarkable skills in architectural technology.

A third idea is that the Bai Yue put the body into the coffin, which they made on top of the crag. They then winched the coffin down from the top and into the cave. This method "from up to down" would have saved a lot of labour. In those times the mountainsides were wooded, and again labour could be saved.

As the Bai Yue increased, from generation to generation, they gradually expanded their territory. Their custom of crag burial spread even to southeast Asia. Their descendants lived in Sichuan, Guangdong, Fujian, Hunan, Yunnan, and Guizhou provinces, and so similar crag burials have been found there.

None of this is agreed upon by all of the experts. We await their further study of the beliefs and customs of the Bai Yue people. Most hopes are pinned on further archaeological discoveries. Others hope, using primitive tools, to be able to test the suggested methods, to create a modern crag burial. A further possibility of study is to compare Bai Yue customs with the customs of other early peoples of other parts of China. Much remains to be done.

Happy Birthday, Fu Kailin!

(Caroline, Wushan, September 23, 1992) A young man feeds the tiny yellow chicks from a trough cut from a water bottle, first grain from one of the polypropylene sacks and then water. The chicks step on each other, crawl into the bottle and get stuck. Gently he separates them and pulls them out of the bottle. Too soon we arrive at the small town that is the turnaround spot for the tourist boats. To my surprise, but not dismay, we stop: a bathroom would be a good idea and lunch is overdue.

Tourist stalls sell onyx carvings at Taiwan prices. Kailin talks a small restaurant into letting us use their bathroom (three flights up) and even into giving us some serviettes for toilet paper. We look for the pears that she says are special in this area, and for snack food. Then all our plans and my dreams come unstuck. The business manager has told Li that the area beyond this point is closed to foreigners and Li asks me what we should do. I say, "Go on, what can they do? Put us in jail overnight? Send us back?"

The answer is all of the above but, worse, they might fine us ¥1000 each. A permit charge. Something. That is convincing. We were far from having that much cash. We have traveler's cheques and you can get money on an American Express Card at certain branches of the Bank of China (on certain days, at certain times) but we have yet to see any, and getting $1000 all at once would be hard. I am still reluctant to give up on cliff burials, but Robert says that he does not think the risk is worth it. He has had a good time here and he is content. It is clear that our Chinese friends would rather not risk a run-in with the police. So that is the end of our adventure.

The tourists are let off at the police boat moored beside the one we had been on this morning. Zhongling talks the boatmen into dropping us at the steps. There he once again refuses to put the army bag on his back. He looks around furiously for a porter but there are none. Silently we trudge, up the ten thousnad stairs, through the smoke of the burning garbage. There, at the top (no fool he), awaits a porter. The "best hotel" is twenty minutes away ¥3. This time even Li hands over her pack.

The porter sets a very brisk pace, but Li is right at his heels. Eventually she and the porter are out of sight. Zhongling also disappears ahead. This walk is pure hell. The burning garbage smell is replaced by gasoline fumes, dust, cars, broken roads and crowds of people making their way home from work. It looks like this town is already moving. With Li and Zhongling out of sight, there is no reason to hurry and so we take it easy. That is fine until there is a junction in our road. Kailin starts asking for the Wushan Hotel. We stop to photograph an old man weaving a fishing net.

We finally arrive at an inauspicious doorway leading to a courtyard. This is a government hotel, full of important officials and foreigners. Well, one foreigner, a German with his Chinese wife, who are travelling with a group from Shanghai. For the first time we are told that our Chinese friends cannot stay with us. There is no room. Robert is offered a room with the German and an important official, and I with the "high status Chinese women." In fact, the German's wife kindly offers to give me her place in a room with the staff so I can have access to the bathroom attached.

These are absolutely the only rooms they have? What are we to do? Take a chance on being let into the "Chinese" hotel? We have to decide on the spot or these places will be gone. Our friends shake their heads. It is not a matter of loyalty and friendship, any more than it was with the Shibaozhai engineers or the second class lounge. We will probably need the amenities of this hotel to make our arrangements. While we wait, the desk clerk comes up with three fourth-class tickets to Yichang tomorrow morning at 6:20 a.m. She claims the boat is sold out. Three are not enough, but we buy them anyway. If we cannot get more, perhaps we can buy two fifth-class tickets and share the bunks.

I have two roommates. One is an older woman who seems to take in washing. Otherwise, she knits. The bathroom is monopolized by her scrubbing and sloshing so long that I am forced by nature to wrap on the door. The laundry not left in the bathroom is hung around her bed. My other roommate is a young and smartly-dressed woman with a male visitor who is stripped down to his blue undershirt. His attention span seems short. As I wait, interminably, to be let into the bathroom, he flips through the TV channels, plays a hand held video game and chats with the girlfriend. There is a large open bottle of what I take to be strong spirits on the TV. What am I to do? Take purse, video camera and whatnot into the swimmingly wet bathroom with me? Or just leave everything on my bed and hope? I stow the cameras inconspicuously, and hope. Which towel is mine? All are wet and gray. The tap comes off in my hand but the warm water feels very good. I get in and wet and dried and out of the bathroom in record time.

The cameras are still there, and so is the guy. At 6:00 they all go for supper. This is my chance to clean up the cameras and recharge the video batteries without drawing attention to them. My bed is beside the window and there is a marvelous view of the harbour. (The air conditioner, set under the open window, runs all night. I cannot figure out how to turn it off.)

The courtyard is packed with important officials dining in the open air. The clatter of dishes and talk is deafening. Waiters scurry everywhere, dirty dishes are piled on tables and in corners of the yard. Then suddenly they all rise for one last toast and are gone, leaving immense piles of dishes and a prodigious mess of bones. The power goes out all over town, except for our hotel, which has its own generator. A waitress tells us they cannot serve us for almost an hour; she suggests the nearby Chengdu Restaurant.

The Chengdu Restaurant specializes in hot pot: this will be Robert's chance for this unique Chongqing dining experience! It is also Kailin's birthday. We will celebrate, why not? The restaurant has no *guanxi* with the generator: it is lit with candles. We are its only customers.

Zhongling orders a very modest hot pot: mushrooms, watercress, bean sprouts, pigs blood (it comes congealed in cakes), bloody eels, dark flat mungbean noodles, Spam, and frogs (not frogs' legs but whole frogs, about as big as your hand). The sauce arrives, is tasted and sent back: it is missing some spices and the sesame oil. It returns halfway through the meal, very heavy on garlic but still missing the oil. Everything is "Just so so," to quote Zhongling, but nothing can dampen our good spirits. We sing "Happy Birthday" and Kailin must make a speech. She says how happy she is to have been on this expedition and now everyone must take a turn, saying something nice. Finally the rice arrives.

Hot pot is reputed to have been invented in Chongqing. It was, so the story goes, a cheap and warm dish favoured by porters and trackers. It is a fondue. The hot pot dish is a wok divided into two compartments: yin and yang. One side is for hot sauce, the other, not so hot, *"bu la."* One starts with a *gin* (pound) each of pork and beef fat, a lot of chilies and a special mix of spices in which the numbing *huajiao* is a major ingredient. Real men prefer the hot. Hot pot restaurants have tables with adjustable gas jets.

An authentic hot pot features every part of a pig or cow or sheep's stomach and intestinal track (and there are a lot of parts and textures). A favorite ingredient is pigs' brain, a pink and foamy glob that somehow sticks together. The idea is to drop the raw morsels in the broth, fish them out when they are cooked, and dunk them in sauce. The bones, in the most authentic places, are thrown on the floor. Afterwards everything is mopped down. Pig stomachs were once cheap, hence the trackers' affording them, and the fiery broth will cover any lack of freshness or want of taste. Hot pot restaurants are everywhere. Ironically the stomach parts are expensive these days and Robert is not terribly keen on them, so we forsake authenticity. A yuppie hot pot would have prawns, exotic mushrooms and other imported treats.

The bill is ¥97. We cannot believe it. Li borrows the waitress' abacus and starts, very professionally, to add it up. Every dish, the broth, the sauce, has a price; 97 it is. Disgruntled, we pay. Actually, it is a cheap hot pot. I have been to feasts where expensive dishes kept coming and coming, and we ate until we were sick and silly.

The power comes back on and our little restaurant loses its romantic air. But the lights bring people back out into the streets and all of Wushan passes by. It is like Saturday night in a small town at home. We dump the accumulation of frog and eel bones into the remains of the hot pot, in true Chinese style. Our plastic containers are rinsed with boiling tea water and everything is poured into the gutter. Dishes done, it is back to the hotel.

Zhongling is anxious that we reserve hotel rooms in Yichang, where we will arrive late tomorrow afternoon. The idea of a long and good night's sleep is inviting. I am still unclear as to how we will return and am under the misapprehension that the boat will take only 48 hours. First we call Norman. A business associate from the US will arrive tomorrow, a good reason for me to get back. He has seen Mr. Xu's furniture factory, which is fine, but the amount has risen to $200,000 and Xu cannot seem to say what it will be used for. Equity investment is a mystery here. Norman says come home. I reply that it is not like the Toronto subway where one can get on and off at any station!

It is much more difficult to get through to the first, second and third best hotels in Yichang. It takes all the charm Zhongling can muster to keep the young woman at the desk dialing their numbers. However, they all say that they have no triple rooms and that doubles are FEC ¥200! Three rooms, $150. The phantom of the wonderful night fades.

Kailin describes how she and her mother got on board ship (as it recommends in the guidebooks, in fact) the night before sailing, and slept there overnight. It is becoming increasingly clear that this must be our first choice. Arrive in Yichang tomorrow, buy the boat tickets back, sleep the night on board. If there are no tickets available, well, we will have to trust in Zhang Fei (and our donation of FEC¥ 50) to help us find a cheaper hotel. Zhongling argues that when we finally got through to the Three Gorges Hotel (number three on everybody's list) we got the "bad boss"; tomorrow we might try for another boss.

So that is that. Our friends go back to their hotel and I sit in the lobby, writing. It is far quieter than my room, as it turns out. The man is still there when I finally return; the two of them are watching a rock show on TV. The other woman has her quilt over her head, despite the heat. Finally, the show is over and he leaves. The air conditioner below the open window runs all night.

10

XILING GORGE

In Search of the Xia and the Shang

(Caroline, Wushan, September 24, 1992) No need to worry about waking up in time. At 4:30 an alarm rings. I have just about had it: there was too much garlic in the hot pot, and my head aches from exhaustion. The two women one by one use the bathroom. The man comes back, still in his undershirt, to use it too. By 5:15, they are gone. I finish recharging the video batteries and trudge down five floors, looking futilely for the *furan* to give me back my ¥10 key money. Deserted. Go by Robert's room. The light is on, no need to knock. We all meet for the van that will take us to the docks.

Most boats moor in Wushan overnight, lining up to make the trip through the final two gorges once it is light. Our boat, however, is not in yet. The peddlers are working the crowd. Kailin finds squares of steamed rice cake, sweet and still warm. We eat it as is; there is no water here for coffee. A small, quiet woman crouches beside me, silent, pointing to her pears and crullers. I point to Kailin: she does the buying. Kailin has already judged her wares unworthy, and that is that. A group of old folks are doing *taichi* at the cliff bottom. Young porters sprawl on the sand, also eating rice cakes. There is a pyramid of pottery, packed in straw, waiting to be loaded. People laugh and chat and do their best to find something to eat and to be comfortable. Bit by bit the sun burns off the yellow morning mist. The serenity of waiting for the ferry on Yangzi stairs is a feeling I will not soon forget.

Our boat appears in the distance, docks, and unloads. Then, peasant, worker, intellectual, foreigner, rich and poor, we board. We have only three berths, in fourth class. All Zhongling's wizardry cannot change that, but he will try again in Badong. This time the berths are in the middle of the cabin. We are hemmed in on both sides by smokers including a group of soldiers playing cards. These men are up and down and in and out, but they smoke inside because the wind would blow their cigarettes out.

My headache is worse. I lie down amidst the luggage on the top berth (our other four have two to sit in) but the smoke drives me down. The others go out to see the Wu Gorge. I am tired of scenery, and just plain tired. Caffeine is good for headaches but a cup of instant coffee just jangles my nerves. Lunch, when it comes, is rice with egg and tomato soup on it, soothing for an overloaded stomach.

We stop so long in Badong that I fear the boat is in need of repairs and they will ask us to get off. I am counting the hours to Yichang, the hours until the Foreign Exchange Section of the Bank of China closes. We may need some more money to pay for the boat tickets back and food for five people. In my mind's eye (a smoky mind), I see a vision of white sheets on single beds in the two-person second-class cabins. No smoking. A western-style toilet down the hall . . . a quiet time to rest and write and talk, before facing the problems of Chongqing. A passage back to the world of everyday life. Robert says not to worry: the current is making our trip very fast. When we are moving. Worry? I have not built into my calculations a wait at the locks of the Gezhouba Dam.

From the deck we have a bird's eye view of men with Tujia carrying baskets, standing patiently in line, presenting themselves, leaning on their sticks with the crossbar on top to receive their loads, then standing up with the help of the sticks and setting out. They sing a tuneless song, a ritual of carrying work.

After two hours at Badong, there is no stop at all at Zigui. Xiling Gorge starts with high mountains, the valley opens wide, then high mountains close in once again. A detailed guidebook names five or six gorges. After all, the "Three Gorges" are a human construct. We are still taking turns resting in the berths, having had no luck getting any more at Badong. It is Kailin's turn and Li has long ago disappeared. No berth needed for her, just a ticket to the movies! The dam site, Sandouping, is just before the end of the gorge. Could we come so far and miss it?

A gravelly slope, a rockslide in fact, has three signs in enormous red characters. Do they say "Proposed site of Three Gorges Dam?" Will the dam be anchored to these mountains? (I later learn that they say, "New townsite of X.") Further, our book says the dam will be built at a low place, the bulk of it manmade, built on a solid granite base and reinforced. Robert says the Hoover Dam is built in a narrow place and the natural rock on either side is its weakness. We pass Maoping.

First Kailin, then Li reappear. We will all be on deck for the sail past. Li explains that the dam will be built more or less on Zhongbaodao Island, a large flat island near the town of Sandouping. Its grassy terraces attracted neolithic farmers and the site, found in the 1950s, is one of the most important in the region.

The first systematic excavations were in 1978/9; they resumed in the mid 1980s under the direction of the Yichang Museum and the Department of History of Sichuan University in Chengdu. The idea was to study the site before the construction of the dam. Li took part in the field work in 1986, under Wang Xiaotiang. They found the remains of wooden houses, ground stone tools, pottery and bone needles and other bone tools. She points to the low buildings that were their living quarters. "No one," she says, "lives there now."

Archaeologists have identified three main cultural layers here: Daxi, Qujialing and a culture similar to the late Erlitou cultures elsewhere. Daxi Culture here is divided into four phases. The proportions of painted and stamped red ware, and then of gray and black ware, and of this or that kind of vessel, determine the sequence of development. Most striking are the stamped and painted pots and the pottery balls that some think were used as fish net floats. Both the Daxi and Qujialing cultures grew rice and, even in the early period, rice husks were used as temper for pots. Both cultures drew the legs of the dead up against their chests for burial.

In the later Qujialing Culture (3500-2500 BC, named after a site in Jinshan County, Hubei), the pottery is not so showy: most pots are gray, then black, and the painted decorations give way to a preference for cord designs. Some archaeologist's tend to stress the continuity of Qujialing Culture with Daxi Culture. Li Wenjie (1979(2):161-164) decribes them as two successive cultures with internal connections: the Daxi stone tools are mostly chipped, Qujialing tools are mostly ground or polished. In Daxi Culture large stone *fu*-axes predominate. Daxi had stone *qiu* balls, but few stone arrowheads. (more war? more reliance on hunting?) Qujialing has more arrowheads and the stone *lian*-sickle appears. Painting is reserved for thin "egg-shell" pottery and spindle whorls. However, Meng Huaping published an article in 1992 arguing that the two cultures differ in origin. Daxi Culture was gradually replaced as Qujialing Culture spread westward through the Yangzi Valley.

The pots of the third period show many similarities to the Erlitou-like cultures of the Central Plains, indicating increasing contacts between peoples of both areas. Erlitou Culture marks the beginning of the Bronze Age in China and graves clearly indicate class distinctions.

Until the past fifty years, lists of the dynasties traditionally started with the qualifier "semi-mythical" for the Xia and the Shang. Archaeologists have vied to correlate later written records and legend with their new finds. The Shang Dynasty became historical fact with the excavation of Yinxu and Erligang in the 1930s. Then in 1949, Yin Weizhang identified Erlitou Culture (first found in Yanshi, Henan) with the Xia. Archaeologists now believe that they can demonstrate that its development was "arrested by the appearance of a new culture (Shang) which eventually assimiliated it" (Yin 1987:357). Legend has it that King Jie, the last king of the Xia was overthrown by King Tang of Shang. What was the Xia Dynasty? One of perhaps several "kingdoms" thought to have been unions of walled towns, incipient state societies, the beginnings of complex society in China.

In 1988, the distinguished Chinese archaeologist An Zhimin, in a general survey in *Current Anthropology*, wrote that, "... the middle and lower reaches of the Changjiang (Yangzi) constitute a rather complete system" (1988:754), but the increased pace of research in the succeeding years has expanded his chronology and filled in the gaps at an astounding pace.

Xiling Gorge has been surveyed by the relics bureaus, museums in Hubei, Nanjing University and the National Museum of History. A month of field work in 1985 identified 63 sites in Zigui and Yichang counties, dating from the Neolithic to the Bronze Age. The earliest sites, dating to 5000 BC, are found in the lower layers at Liulinxi, Chaotianzui, Nujiahe, and Woupendun, characterized by rough low-fired cord-marked pottery and stone tools. They are assigned to the Late Chengbeixi Culture, the earliest yet found in the Middle Yangzi, named after Chengbeixi in Hubei.

Daxi Culture (4500-3500 BC) is found at Gongjiadagou, Wuxangmiao, Zhongbaodao, Chaotianzui, Yangjiawan, Qinshuitan, and Sandouping. There are indications of strong connections between Chengbeixi and early Daxi Culture. Daxi is followed by Qujialing Culture (3500-2500 BC). A Longshan-like Culture (2500-1000 BC) is found at Baimiaozhi and Zhongbaodao, where it is characterized by distinctive grey pottery.

The differences in ceramics between these cultures and those of the Central Plains are meaningful only to experts (provided that they can be upheld). Generally, the layman will note that cord-marked pottery of the Early Neolithic (? to 5000 BC) was followed by a remarkable and beautiful painted pottery (Daxi, Yangshao, Middle Neolithic, 5000-2500 BC). The Late Neolithic (2500-1500 BC) was characterized by a thin greyish or black incised pottery (Longshan, Qujialing).

In *Sandai* (Three Dynasty, Xia, Shang and Zhou) Times (c. 2200-1000 BC), bronze fish hooks and "shoe-shaped" knives appear and there are similarities with the Bashu Sanxingdui Culture of the Chengdu Plain. The Ba were considered to have dominated this area in prehistoric times. The Kui State, mentioned in regard to Baidi, is identified in Western Zhou and is sometimes described as a transitional form to Chu Culture. Archaeologists can now trace the movement of Chu cultural remains west along the Yangzi, replacing the Ba and the Kui. Chu lasts into the Warring States Period (Yang 1992). Archaeologists and historians are only starting to be able to correlate with any certainty these cultures with the historical record, or to draw ethnographic parallels to surviving National Minorities in the adjacent areas.

Only after 2500 BC does primitive metallurgy appear. The Bronze Age occurred later in China than in the Middle East, and at one time it was thought to have been imported from there. Now it is believed that it was begun independently in China, but there is no agreement as to where or how. Bronze making is far more complicated than pottery making, and metal winning — mining and smelting — is more difficult than metal cast-

ing, and requires a much more complex social organization.   Ursula
Franklin points out that, in better-known societies, extensive production of
bronze required slave or other enforced labour. The Hubei copper mines
are known much later in the Spring and Autumn period but much work
needs to be done to determine local Shang-period sources of copper and tin
ore.

This leads to the fascinating but vexing question of when the period of
slavery began in China, or whether there ever were slave states similar to
Greece and Rome at all. The ownership of a few slaves by a class which still
seems to take part in productive labour is not yet a society where slaves are
the foundation of the economic system.

We assemble at the rail to be photographed, with the dam site in the
background. This low, gravelly spot is our reason for being here. Zhongling,
tall and handsome in his immaculate white cotton sport shirt, spreads his
arms wide in triumph. Li, in her "African Animals" T-shirt, leans a propri-
etary arm on the rail. This place is hers. Robert and I pose in our Tilley
hats, our various cameras slung about our necks. Kailin must have taken
the pictures because the picture is only of the four of us, without her.

Baleful Badong: Splendid Shennongjia Stream

(Ruth, Badong, October 30, 1992) I began my visit to the Yangzi in Yichang
and then took a ferry up river, first to Badong and then to Zigui. I stood at
the front on the middle deck, snug in my padded jacket. My guide Susan
kept leaving our third-class cabin from time to time to explain points of
interest and tell mythological stories about the whirlpools and Monkey and
Guanyin, the Goddess of Mercy.

Monkey is a character from a Qing Dynasty novel called *Journey to the
West* which, like the *Three Kingdoms*, is a collection of folk tales. It tells of
a naughty animal with supernatural powers much beloved of Chinese children. Monkey helped a monk Xuan Zhang overcome near impossible obstacles to go to India and bring back the Buddhist *sutras*, or scriptures. The
monk passed here on the way to India, Susan said. At that time, the Yangzi
was very turbulent and a thousand-year-old turtle appeared. The monk
crossed the river on the reptile's back.

The monk promised in return to tell the Goddess of Mercy to invite
the turtle back to Heaven. The turtle had once been a god but had been
banished for ungodly behaviour.

At the Yangzi on their way back, the monk and his friends found the
turtle waiting, ready to ferry them back across the river. When the monk
told the turtle he had been too busy in India to remember to ask the Goddess of Mercy to intercede for him, the turtle became so angry, it shook its
tail three times. The passengers fell into the river but were saved by Mon-

key. The Buddhist scriptures became wet and were put to dry on a stone now called Scriptures Sunning Platform. The turtle's three shakes resulted in three whirlpools that endangered ships. (The sites have now been blown up and destroyed.)

Susan had been a university teacher of English in Fuzhou for two years, living in worse conditions than when she was a student. She was ambitious. She did not want to make only ¥350 a month plus subsidized housing. She and her fiance had to borrow a large sum of money to pay off the school so they could move back to her home in Yichang. The government had paid for her education and she was obliged to work for the university in return. On a trip back home, she had applied for a job she happened to see advertised on television with the travel agency. Fortunately, it was close to her parent's home. Hired three months before, she had been given no training on how to be a guide. Training was only once a year and there had been no time, she said.

The Xiling Gorge was indeed striking, rising high on either side of the narrowed river up to 900m. We passed Sandouping, eight kilometres east of the site of the new dam. It had three new white modern buildings from 1984. These had been put up for the Three Gorges Dam project when it was first announced, when the Canadian government paid for a feasibility study. The buildings were never used because the decision for the project was postponed in 1989 for five years. A decision to go ahead was made in 1992.

Flat Zhongbaodao Island was busy with thousands of curious Chinese tourists because it was the dam site and would soon disappear. Susan said there was nothing to be seen. The island has 32 archeological sites with graves from the 21st century BC to the 13th century AD.

We passed the doomed towns of Maoping and Xintan. There was also a town partially destroyed by a 1985 landslide, which is now a bare hillside sloping into the river. The site was marked for the relocation of one of the fated cities.

We passed Xiangxi, the port for Wang Zhaojun's home town, where the clear Fragrant Stream met the muddy Yangzi at a stark white statue of that Chinese beauty who married a tribal emperor two thousand years ago. A guide once told me that the water was clear because Wang Zhaojun dropped her pearls there. We sailed by Zigui on the north side, a sizeable town four hours from Yichang. It had a row of Tujia men backpacking coal in their distinctive baskets from a hill to a truck.

There were long lineups in the ship's canteen and we tried three times unsuccessfully to get food. "I need good food every day or I get sick," Susan said. I offered her some of my own food. She ate two of my few remaining granola bars and said it was not enough, and went off to try again. Some nice people in our cabin give me a pomelo. Back on deck, I was approached by Mr. Hsu, a boyish-looking young man in a brown corduroy jacket going

to check out a cement plant in Badong. It would be one of the many supply-
ing eleven million tonnes of cement for the dam, and for the thousands of
new buildings.

Hsu spoke good English, and it was obvious he enjoyed his contact
with the outside world. He asked if I knew "Moon River," one of my
favourite songs. Like everyone else, he said the advantages of the new project outweighed the disadvantages. Susan complained that it was too cold
for her on deck and stayed in our room, warmed by the crowds of people on
the eight bunkbeds. The ship had no first or second class. Lower-class passengers huddled in the passageways, on the decks, anywhere there was
space and warmth. A man on the deck below us tied a string of brilliant red
chili peppers onto the mast to dry in the sun, a colourful touch.

I preferred the outdoors. The scenery was beautiful and Hsu was
amusing. He tried to sing "Scarborough Fair," but neither of us knew all
the words.

Badong was on the south side of the river, very hard to photograph in
the afternoon sun. It was not as big as Yichang and would be flooded up to
80m above its current waterline. The whole town of thirty thousand would
move five miles away over the next ten years; 5,800 people had already been
relocated, arranged by their work units. They were reassigned new apartments; they did not get allowances. The evacuation would not be Badong's
first. The town had already changed locations twice since the Northern
Song dynasty in the 10th or 11th century. It was formerly on the north
bank, the switch intended to give it better transportation links with towns
to the south.

I remembered visiting Iroquois in the 1950s. This eastern Ontario
town had to be moved to make way for the St. Lawrence Seaway. Complete
houses were carried by trucks along the highways. But in China's towns,
most people live in apartment buildings. Their housing would be rebuilt at
the new site and they only had to move their furniture.

The port was full of porters, Tujia tribespeople carrying baskets on
their backs and T-shaped walking sticks with which to prop them up while
resting. We hired one for ¥1.50 (CDN$0.33) for the climb to the govern-
ment guesthouse, where I was put into the best room, no other being available. It was a large two-room suite with private bath, broken toilet seat, no
towels, thin curtains, dim lights, and wet floor. There was no place to
change money, no international direct dial. Still, it was better than I expected.

Mr. Tian Yuan, former director of relics and former director of
tourism, took us to the first historical relic on my list, the Autumn Wind
Pavilion high on the hill above the city of Badong. He too said the dam was
the government's decision and the benefits outweigh the disadvantages.
The ten metre-high pavilion has two stories and is in good condition. It had
originally been built in the Southern Song dynasty by Kou Zhun, Governor

of Badong county, so he could meet with the masses. The building had been destroyed three times by war and fire and was said by Mr. Tian to have been last rebuilt in Song architectural style in Guangxu's reign, 1875-1908. The Autumn Wind Pavilion was listed in our Canadian feasibility study as being at an elevation of 175m and as having been built in 1816-1860.

The pavilion was repaired in 1992 and still had the original Qing tiles and curled dragon-tail roof ends. The round-bottomed, overturned wine cups on the roof originally had straight bottoms. They made me wonder if something more than gubernatorial business went on here! Wooden pearls used to be in all the dragons' mouths in the upturned corners of the roofs, and bells used to tinkle in the wind. No longer are there the many tablets with their eighty poems, praising Kou Zhun. These were destroyed between the Song and Qing dynasties. This pavilion is valued because of Kou Zhun, a "kind and fair magistrate," a great man "because he listened to the common people," according to Tian Yuan.

Kou Zhun's pavilion is near the top of the hill, the town line. When the dam is finished, it will still be 30m above the water line, but it will be moved along with the town. The pavilion is currently in the middle of an army camp but, after some renovations, the army will move away and it will be accessible to tourists until the relocation.

There used to be other historic buildings in this area, said Mr. Tian, but they were destroyed by Japanese bombers. A Tang dynasty Temple of Longevity that was never rebuilt.

For the other buildings on my list, graves up to 2,300 years old, Mr. Tian said I would have to go with him tomorrow. It was the peak of the tourist season; 500 tourists would be going to a Tujia village and taking boats down the Shennongjia Stream, and he had to help. I would have to pay ¥250 for the two of us as my prepaid trip did not cover it, said Susan.

I explained to Susan that I did not have that much money with me, and she managed to haggle the price down to ¥120 for two. Then she said that she had told me to bring ¥2,000, (US$364 at 5.4), a fantastic sum for one who had prepaid a six-day trip. I wondered if this was a scam to get me to pay for a sightseeing trip I did not really need? But my curiosity defused my scepticism.

Susan complained that her room (¥15) was small, without a bath, that my room was more expensive (¥90), and that the money she was allowed for both (¥70 a day) was not enough. I argued that her agency should have checked the price before we left and she mumbled something about being in another county. Somehow all worked out, and she shared my bathroom rather than the filthy public rooms.

The guesthouse had no eating facilities. Mr. Tian recommended a fourtable cubbyhole around the corner named Rendafuwubu, People's Congress Restaurant. It was an overly ambitious name. The waitress was kind enough to scald our dishes and chopsticks. The food was so good, in spite of the cracked and chipped plates, that we kept going back to eat there. The restaurant was privately run. This in itself was usually a sure sign that the service was better than a government-operated establishment. The women had raised the money themselves from relatives. They were not bitter about the move to the new town. They did not have much to move and could operate just as easily in the new town, said Susan, who refused to translate because she was tired.

Badong looked generally old, as in rundown, not antique. The main commercial street running east-west was flat and partway up the mountain. A lot of trucks and buses roared noisily through it. Staircases joined parallel streets above and below the stores. Shops were well stocked, but not with luxury items. Only a few peddlers stood outside, each selling a miniscule amount of fruit, combs, or clothing. This was not Beijing. But there were a couple of new-looking buildings and the people were friendly and curious.

A rooster started crowing at 2:00 a.m. A ship's horn blared at 2:42 for 15 seconds. We got up at 5:00 a.m. to find the restaurant open but not ready to serve food, as promised the night before. So we bought some dry biscuits and some hard, reheated dumplings. We finally took off with Mr. Tian in a rickety old bus, up a hill and down to an old ferry pier on the Yangzi in the misty pink morning light. We waited for six trucks full of coal to cross the river on the ferry ahead of us.

Susan said she did not want to come to Badong again. It was her first time to this town. The hotel would not give her a receipt, so she could not get a refund from her agency. She had said the same thing about the food on the boat.

The Tujia village was 47km from the ferry pier on the north side on Shennongjia Mountain. Further north, a huge, hairy humanoid is said to have abducted a woman in 1976, Shennongjia's own Sasquatch. The wild one had been sighted by a couple hundred people on separate occasions in the last thirty years, said a tourism magazine. Scientific expeditions had been sent to investigate but nothing is related of what happened. We saw only vast, beautiful forests and a few villages.

We drove through the mountains along dusty, unpaved roads with scarves over our noses. Susan said I was asking too many questions. She was tired. She repeated for the third day that she had not slept the night before. She was sick, she explained, and I wondered why she was working.

The villagers were ready with sedan chairs to carry tourists down to the stream. Women all along the stone pathway tried to sell us grapefruit-like pomelo and peanuts. Shy women and some men were carrying babies in the inverted "L"-shaped baskets on their backs, an unusual design. We stepped into one of the many 18-seat canoes with Mr. Tian.

The ride down the stream made me appreciate what an opportunity I had been given. It was stunningly beautiful, breathtaking, the clear shallow rushing stream at times only six or seven metres wide, the sheer limestone cliffs rising almost a kilometre beside us. The scenery was, in places, green and pristine, the earth red. Here and there were willowy bamboo and patchwork-quilt hillsides and terraces of potatoes, wheat and cotton.

A boatman in front steered with a paddle. Around the many bends, three others would jump from the stern into the water, their feet shod only with cloth and straw sandals. They strained mightily so they could maneouvre us safely around the sharp corners. The only people we saw were other tourists far ahead of us, or in the restaurant where we stopped for lunch halfway through the trip.

And then we came upon the boatmen, three of them harnessed in leather across one shoulder as they pulled the huge canoes upstream. I had read of boatmen like these in history books. They were shouting something like *"wei, wei, li, hou,"* to deaden the pain. No more are these men seen or heard on the Yangzi itself. Soon these would be gone too, and I could not blame them for wanting a better, easier life.

The terraced hills made me wonder just where the million people would be relocated. Crops were planted to the very top, wherever there was relatively level ground.

Mr. Tian talked about the mysterious hanging coffins in the caves in the cliffs but I could not see any. I did see square holes about 11cm each side, cut neatly in the walls. Mr. Tian said no one knew what they were. And he mentioned the three hundred different medicinal herbs while we looked for wild monkeys. He scolded a boatman for throwing pomelo peelings into the clean, clear water. The boat tours on the Shennongjia Stream had been his idea, his baby. He seemed to enjoy sharing his exquisite find with us.

We arrived all too soon, three and a half hours from the start, at the mouth of the stream near Badong. Suddenly I realized that most of this beauty, most of this clear, lovely water would be muddy brown when the dam was built and the Yangzi backs up here 25km. The new dam would kill the drama of the cliffs. But who knows? Maybe the rich silt might make land here more fertile. But what if it were polluted with chemicals? Maybe the higher water level would give better access to other beautiful streams further upsteam, closer to the land of the the hairy wild people.

We switched to a ferry that took us back to town. Mr. Tian pointed to places at the mouth of the stream and along the shore. "There are the Spring and Autumn Period, the Song and Han dynasty graves." As we parted, I asked Mr. Tian what he would do if he were given ¥20,000 (about US$3,800). I expected him to talk about saving relics.

"I would help common people get rich," he answered quickly. "I would not spend it on myself."

Badong was dirty, dusty, and polluted with coal dust and truck fumes. Pollution control is a luxury of the rich. I walked alone along the main street, up the hill to the edge of town where I could see the river. There was no place along the road to sit, no park, no place inviting. People smiled shyly but no one spoke to me.

At dinner, Susan explained she had a chronic pain at the back of her head that kept her awake. She took no pills because she did not want to become addicted and she could not afford the massage I suggested. I told her again that I was writing a book and had to make inquiries.

She did not like my questions, she said. She sometimes answered them, instead of asking a person like Mr. Tian. I explained that Mr. Tian was an authority, to be quoted by name, not her. She apologized, but later continued to do it. I had to coax her to translate for me.

I told her I would give her ¥80 for a massage, because I felt it might help her get better and health was very important. As she spoke I kept thinking of the story I had heard about other guides who told each new tour group, "Today is my birthday, but because you needed a guide, I decided to work instead to help you out."

Recollections of the Old River

(Caroline, Xiling Gorge, September 19, 1992) Isabella Bird met Norman's grandfather, the Very Reverend James Endicott, on the steamer *Chang-wo*, from Wuhan to Yichang in 1897. He was returning to Sichuan with, we assume from the date, Norman's father's older sister Enid ("the inevitable baby"), Norman's great grandmother and another couple. They had fled the antiforeign riots of 1995. Isabella's cabin was directly over the boiler. They very kindly offered to relinquish the ship's "saloon" and she gratefully accepted. There, in relative coolness, she could print her photographs: the genial engineer providing distilled water. From the "Chinese cabins" came the faint, sickly smell of opium smoke.

She was grateful for the Canadians' companionship and good advice: they had been living in the area she intended to explore. When, like so many feckless travelers, she asked Rev. Endicott about sights on the ascent of the Gorges, he replied: "People have enough to do looking after their lives."

After going up through the Gorges, and coming back down, Bird concluded: "I have found that many of the deterrent perils which are arrayed before the eyes of travelers about to begin a journey are greatly exaggerated . . . . Not so the perils of the Yangtze . . . . The risks are many and serious and cannot be provided against by any forethought."

She describes the Hsin-tan Rapids.

The boatmen turned us and our servants out at 10 a.m., and we stood about and sat on the great boulders on the bleak mountainside in a bitterly cold, sunless wind each day till nearly five, deluded into the belief that our boat would move. A repulsive and ceaseless crowd of men and boys stood above, below, and behind us, though our position was strategically chosen. Mud was thrown and stuck; foul and bad names were used all day by successive crowds. I am hardened to most things, but the odour of that crowd made me uncomfortable. More than 1200 trackers, men and boys, notoriously the roughest class in China, were living in mat huts on the hillside, with all their foul and ofttimes vicious accessories.

A glorious sight the Hsin-tan is, as seen from our point of vantage, half-way up the last cataract, a hill of raging water with a white waterfall at the top, sharp, black rocks pushing their vicious heads through the foam, and above, absolute calm. I never saw such exciting water scenes — the wild rush of the cataract; the great junks hauled up the channel on the north side by 400 men each, hanging trembling in the surges, or, as in one case, from a tow-rope breaking, spinning down the cataract at tremendous speed into frightful perils; while others, after a last tremendous effort, entered into the peace of the upper waters.

There were big junks with masts lashed on their sides, bound downwards, and their passage was more exciting than all else. They come broadside on down the smooth slope of water above, then make the leap bow on, fifty, eighty, even a hundred rowers at the oars and yulows, standing facing forwards, and with shrieks and yells pulling for their lives. The plunge comes; the bow and fore part of the deck are lost in foam and spray, emerging but to be lost again as they flash by, then turning round and round, mere playthings of the cataract, but by skill and effort got bow on again in time to take the lesser rapid below (1899:119-20).

The tremendous crash and roar of the cataract, above which the yells and shouts of hundreds of straining trackers are heard, mingled with the ceaseless beating of drums and gongs, some as signals, others to frighten evil spirits, make up a pandemonium which can never be forgotten (1899:123).

Not all of the excitement came from Nature. Norman descended and returned through the Gorges four times, 1927-28 and 1933-34 on furlough. In June 1941, the family had to fly from Chongqing to Hong Kong, by DC3, because the Japanese had occupied the lower Yangzi. Many of the seats were full of bales of pig bristles, a valuable export item in the days of shaving brushes.

The first time, Norman was two years old and does not remember, but a tiny incident sticks in family lore. The Northern Expedition to reunite China was under way. The Wanxian Incident caused an outbreak of nationalism and antiforeign feeling and the missionaries were advised to flee. His father, James, did not want to go but his mother, Mary, was pregnant with his brother Steve. There were bandits along the river; huge steel plates had been bolted to the railings of the steamships. Foreigners were invited up to the bridge, which was even more strongly fortified. On one occasion there was a lot of firing and when the family returned to the stateroom, they found that a bullet had pierced the side of Norman's potty. This was a good story for raising money in Canada.

In early spring 1933, they went down again, on a Jardine Mattheson steamboat boat called the *Kungwo*. (There were two British steamboat lines, J. M. and Butterfield and Swire.) The *Kungwo* had sailed from a pontoon landing just below Huang Family Lane, in Chongqing, an area called Huangjinmiao, for a temple that long ago burned down. Norman, who was seven years old, remembers a ceremony that involved cutting a chicken's throat and spilling its blood on the bow. The boat also had two eyes painted on it, to spy out the dangerous rocks.

At Goosetail Rock, in Xiling Gorge, the helmsmen did not make the sharp right turn in time, and the boat hit the rocks, making a leak between the plates. The first class passengers had been assembled on the bridge for the last leg of the journey into Yichang that night. Norman remembers that the sun was getting low, and that you could feel the bump, and someone saying that that was unusual on a steamboat. It began to list, there was shouting, and the captain headed full steam ahead to shore, beaching at an angle of twenty degrees on a sandbar. Even with a very small crack, a boat takes on a lot of water very quickly.

There were great comings and goings and vehement discussions about bandits and warlords. The captain issued rifles to the foreign men and they posted guards all night long. In the morning, the captain had great difficulty contracting some large junks to take the passengers and cargo down to Yichang. With oars and sails, it took a whole day. And so the Endicott children travelled same way their grandfather had, forty years earlier.

Swollen with our own sense of adventure, we tend to forget how much has changed over the century. In fact, the river Isabella Bird, Norman's grandfather, and Norman himself, travelled is no more.

Zigui and the Qu Yuan Memorial

(Ruth, Badong, October 30, 1992) The next morning, we went to buy our boat tickets for Zigui, one hour downstream from Badong. The 10:30 ship had been cancelled because of mist at Gezhouba, and Susan went looking

for an alternative. The tourist office phoned the other three boat companies and found a ship leaving immediately. We ran across town and down the steps to catch it. It would have taken three and a half hours by bus otherwise, or a wait for the afternoon ship. This time we had a bed in a fourth-class cabin with eleven other beds. It cost us ¥7, a little over a dollar.

The ship stopped at Xiangxi, eight kilometres east of Zigui. It did not seem to have a wharf and we all had to walk a wobbly, narrow gangplank. We caught a minibus. Zigui only had one wharf, and it was crowded with two of the fancy tour boats, one of them from Susan's company. We got off where the ships are docked and the manager of the travel service took us to lunch, a chicken hot pot where we cooked the meat, vegetables and noodles ourselves in soup at the table. It was delicious.

Though it was one of the poorest towns in China, with a total industrial output of ¥387 million in 1991, Zigui looked newer and wealthier than Badong. Its roads were better paved, with fewer holes in the sidewalks. The buildings were better and stronger. The market was bigger, more prosperous looking.

I walked by a few peddlers and one of them asked me to take a photo of a man selling pots and pans, with a kit for mending pots. He looked quite unusual so I clicked my camera. "You'll have to pay him five yuan," said the peddler. I was taken aback but I countered with "No, he has to pay me ten yuan for taking *his* picture." The first peddler backed off and I continued up the street expecting someone to shout at me, but no one did. This town is out to make a buck, I thought uncomfortably, but I was wrong. Another peddler who had been within earshot was laughing.

When I wandered down the street to a group of seniors playing croquet and checkers, the atmosphere was so friendly I had to change my mind. It was like other Chinese towns I had visited. People invited me to play. They tried to talk to me, though I could not understand anything but their warmth. The referee for croquet was so officiously funny, I could not help laughing.

Down another street, I came across a billiard game where again I was invited to play. Then, when I climbed up the mountain, I bumped into our bus driver, who remembered me. A friendly lady named Xiang, with a warm smile took me up the mountain to see the orange trees and a view of the river from the heights. And the new school — new buildings in a town that would be destroyed? She sounded pleased when she said she was moving to Yichang. But with no Susan, I could not ask for details.

Later Miss Xiang showed up at the hotel with some oranges for me and some Mao buttons for sale. She did not seem disappointed when I said no to the Mao buttons. I loved her for it; I would have done the same in her position. I took her name and address to send her the photo we had taken together and felt that she would survive the move.

Back near the hotel, the first peddler asked me to take a picture of him and his daughter. No charge this time. We laughed at our joke. I bought some sweet Chinese gooseberries, tiny kiwi fruit for 80 *fen*. Then I went looking for the blind man seen earlier. I wanted to give him something in return for the happy time I had had in his town. But I discovered he had a thriving business telling fortunes and did not need help.

I liked Zigui.

Tourists go to 4,000-year-old Zigui because it is the hometown of Qu Yuan, the man in whose memory dragon boats are raced once a year. The city is named after an incident between Qu Yuan and his sister. There tourists could race each other in the dragon-shaped boats and visit the temple to Qu Yuan, paying homage to this patriotic poet-statesman who drowned himself 2,300 years ago in despair at the defeat of the Chu Kingdom. This was the tragedy he had tried unsuccessfully to avert.

In the Three Kingdoms period, Liu Bei built a stone city here. Unlike Badong, it had always been on the same site, though it had moved uphill. Zigui would have its lowest 100m flooded. The town already had a new site 30km away, near the Three Gorges Dam, near Maoping on the Yangzi.

When we visited Qu Yuan's temple, I was startled to learn that the temple was only a few years old, having been moved from its original site because of the Gezhouba Dam. How important was it to relocate a new building? But officials assured me that people wanted to pay their respects to this great man. Tourism was one of the main industries in this town. I watched several busloads of Japanese, Overseas Chinese and local Chinese tourists from the two tour boats at the dock. Could we in all honesty ask people to help save a temple of only twenty years?

I thought of the Cultural Revolution over thirty years before, when relics like this were not only neglected but actually destroyed. History is full of shrines and memorials eliminated by lightning, fires, and wars — Tang buildings rebuilt in the Song, then again in the Ming, only to be destroyed by Japanese bombs. Then the dam. It was not their fault that the current temple was so young. In 200 years, this temple would be 200 years old. Ancient. Should one plant it again for the antique lovers of the future?

I love the colour and excitement and drums of dragon boat races. We have this bit of Chinese culture every year in Toronto. And there are dragon boat teams in Singapore, Hong Kong, the U.S., Australia, Japan, Germany, the U.K. and Singapore. Qu Yuan belongs to the world. His temple should be saved.

Qu Yuan was known primarily as a poet, a writer. He lived from 338-278 BC during the Zhou-Warring States period, about five hundred years before the Three Kingdoms. (Other sources date him about 343-289 BC). He was a son of a noble family of Chu, a state that straddled the Yangzi to the East China Sea. Most of Chu's people were not Han, and thus were considered barbarian by its neighbours. They were of Man origins, a group related to the Miao-Yao.

Chu was a prosperous state because of the abundance of water. Its capital was at Jiangling, near the present town of Shashi. It started out roughly within the borders of today's Hubei province. The civilization of Chu rose during the first millennium. This was the era which developed the technology for great bronze ceremonial vessels and particularly the magnificent two-tone ritual bells. The 65 bells unearthed recently outside Wuhan ranged down from 1.5m high. The molten metal had been poured into molds of such an exact size as to produce five octaves of a twelve-tone scale when struck. The ability to do this has not been found anywhere else in the world. Chu started expanding northward in 771 BC, but was kept in check by the Kingdom of Wu, its eastern neighbour, in the seventh century BC. In the third century BC, the states of Chu, Qi and Qin absorbed the remaining states and fought among themselves for dominance.

As a minister to King Huai Wang of Chu (329-299 BC), Qu Yuan was in charge of drafting government decrees, diplomacy and the imperial family records. He tried, unsuccessfully, to warn the monarch and his son about the expansionism of the Qin. According to his own account, he was maligned by a competitor in the court. He lost favour with the emperor and was banished to the "Siberia" of Chu in the Yuanxiang River valley, where he poured his bitterness and depression into his powerful poetry. His poems emphasized his own sterling qualities compared to the corruptness of his colleagues.

Perhaps he was ignored because the king was so obsessed with projects like the bells, a gift from the King of Chu to a nobleman. Qu Yuan turned out to be correct. In 297 BC (other sources say 278 BC), Qin Shihuang conquered Chu and went on to become the first emperor of China. Shortly after, on the fifth day of the fifth lunar month, Qu Yuan committed suicide by drowning in the Miluo River, a tributary of the Yangzi near Yueyang. Chinese scholars quarrel over whether or not Qu Yuan should have committed suicide. Was his death a justified political statement, or a cop-out?

In the history of Chinese poetry, *the Chuci, Elegies of Chu*, formed the second great school. These are said to contain Qu Yuan's poetry and that of his followers. The poems are in a different metre from the first school, with great flights of fancy and magic (perhaps drug related). Qu Yuan's style and themes influenced many later poets. Scholars have debated whether or not these poems are actually his.

Some of Qu Yuan's poems are over 350 lines long. Many are full of allegories. Qu became famous especially in the Han. He is regarded as the loyal minister who is unjustly rejected by his ruler. This quotation from the "autobiographical" *Li Sao* (Encountering Sorrow) was translated by David Hawkes.

The conspirators steal their heedless pleasures;

Their road is dark and leads to danger.
What do I care of the peril to myself?
I fear only the wreck of my lord's carriage.
I hastened to his side in attendance
To lead him in the steps of the ancient kings,
But the Fragrant One would not look into my heart;
Instead, heeding slander, he turned on me in rage.

In honour of his loyalty, his persistence, his brilliant literary style, and perhaps to keep the spirit of this visionary happy, every year dragon boats are raced on the fifth day of the fifth lunar month, the anniversary of his death. Some say it is to feed the fish before they can eat him up; others say it is to feed the spirit of Qu Yuan, or to look for his body. Though I have never seen anyone throw food into the water during the dozen or so races I have witnessed in China, Hong Kong and Canada, sticky rice, wrapped in bamboo leaves are eaten as part of the festival.

I wondered if the dragon boat races were actually related to earlier festivals and beliefs, tied in with virgin sacrifices to the god of the Yangzi? No one today seems to look at Qu Yuan as a model, though some students might have identified with him as they heard of the deaths of their colleagues in Tiananmen Square. His was an example of resistence to unquestioned Confucian obedience to those above you.

I was attracted to the recently discovered old graves near Qu Yuan's temple. They had been unearthed earlier this year when a new road to the temple had been built. But they were not yet on my list of ancient relics to be saved. Were more graves still to be discovered? How many people lived and died in Zigui in four thousand years? Should we, could we look for all of them? Of course not.

The Significance Of Qu Yuan

(Robert, Chongqing, Sept. 30, 1992) It is the kind of room setting I have seen many times on news clips and magazine pages: Mao Zedong receiving dignitaries; Pierre Trudeau talking to Zhou Enlai; Richard Nixon in his moment of glory. The high-ceilinged room is about ten by twenty metres. In the centre there is an open square of austere armchairs with machine-lace antimacassars over their backs. In between the pairs of chairs, low tables are set with covered teacups. On the walls behind, banners bearing Chinese characters proclaiming whatever.

And now here I am, feeling like I have from the first moment I set foot in China, like a very special guest. The major difference between the news pictures of the scene and the actual experience is the heady aroma of the jasmine tea. The subject of discussion is to be the ancient poet Qu Yuan (BC 342-278). I will learn a number of facts and sense a significant attitude that

I wish I had understood before the trip down the river. It helps to explain a lot of things.

The place is the Chongqing Teachers' College. We have arrived back in Chongqing a couple of days ago. I am totally exhausted and rather sick but there are still a number of things to do. Kailin had asked me to visit her college and give a lecture to her English class. At the last minute, however, it seems that her superiors have gotten nervous about not having the right papers and permissions to allow a foreigner to speak. As we find over and over in China the memory of the Cultural Revolution is a very real and powerful factor. No one wants this week's courtesy or friendly gesture to become a gross indiscretion as judged by next week's standards. Frankly, that is all right by me since I am not really up to much of a performance. So this visit and lunch with some of her scholar colleagues is enough for me.

Kailin came to the hotel to pick me up early. I wasn't sure why we needed to leave when we did. We ran out into the busy street in the front of the Chongqing guesthouse, where she spotted and flagged down the appropriate mini-bus from among the turmoil of jostling vehicles. On board we perched on the sides of the engine casing beside the driver. The reason for the early start slowly unfolded. The trip took almost an hour. This, I realized, was what the average Chinese person experiences just to get to work in a city of four million souls. Kailin had already made the trip once in the opposite direction and would make it twice more by the time she brought me back to the hotel and returned to her room at the college.

The campus of the Teachers' College is laid out in the typical international institutional style. There is a gate, shrub-lined walkways, a lotus pool and a series of four- or five-storey concrete buildings. The student residences are crowded, with four students sharing a room that feels small to one Canadian. Each professor has a bed sitting room with a small kitchen and a toilet/shower room. These rooms are quite practical and comfortable, but meagre in size and furnishing by Western standards. Kailin is apologetic about her humble lodgings but I assured her I am honoured to be there and looking forward to meeting her friends. She reminds me that she also has her family apartment at the university, where her husband teaches engineering. I never meet him, and I wonder if many Chinese professional people live these sorts of semi-separate lives? By now she is almost late for her class.

"I don't want to be criticized," she says.

"It sounds like the Cultural Revolution," I reply. "Would you really be publicly criticized for being late for teaching?"

"There are some people who find it hard to give up the old ways," she says. She introduces me to an eager young colleague, Luo Xingbing, who speaks Voice of America English, instead of Kailin's version of BBC World Service. Then she disappears to her class and leaves me in the capable hands of Xingbing.

So here I sit in the formal meeting room of the college, tea being poured with considerable ceremony. We have been joined by: Professor Huang, the Director of the Qu Yuan Institute; Professor Xuang, Editor of the *Magazine for the Study of Qu Yuan*; and Mr. Guan, the curator of the college's museum. There is also a reporter from the school's newspaper and a photographer who makes us all feel important by continually snapping shots. It is through this meeting that I begin to come to a better appreciation of just how important the study of the ancient poets is to people of the educated class in China. The fact that they speak with animated passion about a poet who lived and worked over two thousand years ago speaks to the continuity of Chinese culture, and perhaps to the recurrence of its themes. The fact that one of these scholars has spent his career collecting the hundreds of poems and verses that relate to the Three Gorges indicates the importance of the written record in Chinese culture. The fact that the whole discussion is, in a way, about something other than the obvious, exemplifies that Chinese quality that Westerners sometimes describe as "inscrutability."

Professor Huang begins: "Chinese traditional culture is divided into two branches, the Yellow River, or northern, and the Yangzi River or southern. In the beginning, culture was different in the north and south. The Yangzi culture represented the local or informal tradition."

"Is it somewhat like what we in the West might see as the difference between high and low culture, or classical, as distinct from romantic or folk tradition?" I ask.

"Yes," Professor Xuang continues. "Before the seventh or eighth century BC the differences were striking. Qu Yuan, as you know, lived in Zigui on the Yangzi River. His work represents the mixing of the two branches. His time was critical in the turning-point in Chinese development. He created unity from the two earlier traditions. The Three Gorges area is in the centre of the country and it is the land that represents the mixing of the north and south."

"The oldest record of Qu Yuan's poetry comes from Han times, perhaps three hundred years after his death," Professor Huang picks up the narrative. "Before his work was recorded by historians like Si Maqian he was known only in oral tradition. All known written records before 221 BC were destroyed by the Qin Emperor at the end of the Warring States Period. Now there are Japanese scholars who claim that Qu Yuan was a mythical and not an actual person. It is important to us to prove the actual existence of Qu Yuan."

"You understand that I am here to explore the possible cultural impacts of the flooding above the Three Gorges Dam," I explain. "Is it possible that earlier versions of Qu Yuan's work might be found in some archaeological sites in the future?"

"Writing was sometimes done on bamboo tablets in ancient times," one of the professors says. "Such tablets have been found in tombs."

"Might there be tombs from the early Han Period in the Zigui area?" I continue.

"Many relics have been found near Zigui."

At this point they bring out a number of books and a scroll with a copy of the Ming Dynasty painting of Qu Yuan. What I have already read about the poet was that he served as an advisor to the king of one of the domains in the Warring States Period. He tried unsuccessfully to warn his master about the encroachments of a neighbouring state. The king didn't like the advice and banished Qu Yuan to the countryside. In the end his predictions proved right and his country was conquered. But Qu Yuan was not vindicated; he was broken-hearted, and committed suicide by jumping into the river. It is hard, therefore, to ignore the imagery of the faithful advisor with bad news for the government.

We look at the picture. "Why," I ask, "is he shown wearing a sword?"

Professor Huang says, "It was a symbol that he was discontented with corruption in government."

There is a long pause. "Then he was a revolutionary," I venture.

Professor Xuang stands up immediately when the translation is complete. "I better go and check on lunch," he says nervously and disappears out the door. Mr. Guan seems a bit shaky too, and gets up to pour some more tea. The discussion is over. Only Professor Huang is calm. He gives me some special packages of tea as a gift. He seems content that I had gotten the message. He wonders if I would like to become member of their Institute for the Study of the Three Gorges. I say that I am like a child in my knowledge compared to them and not worthy to be an equal. But if there is anything I can do to assist them from abroad I will be more than happy to do what I can. I am ushered to lunch, where Kailin joins us.

Wang Zhaojun: Bride of the Hun

(Ruth, Xinshan, Zigui, November 1, 1992) The next morning, Susan turned the light on at 6:30 a.m. and said we were catching a 7:00 a.m. bus across the street. She had not told me this before, and I felt at her mercy. The hotel restaurant was not yet open and we ate more crackers for breakfast, crackers I had bought myself. We waited until 7:41 a.m. for the bus. Susan needed hot food. Her stomach felt uncomfortable when she ate cold food. I suggested she get some hot water to drink from the bus station but she glared at me because there was not any ready yet. She was tense again. She was now answering some of my questions with a bitter, "I told you that yesterday." I scolded her about her rudeness. "I have a temper at home, too," was all she said.

The rickety bus chugged past mudbrick houses hung with batches of drying corn. Even the small towns had modern brick buildings four storeys tall, all part of Zigui "City," though they were obviously separate communities. A snappy song about Mao Zedong played on the bus radio and Susan said it was because people still loved and missed him. We followed the Fragrant Stream north, past swaying suspension bridges, more orange trees, cabbage and sweet potato patches and black pigs. The stream was pristine, clear for miles. Wang Zhaojun had a long strand of pearls. Brown Yangzi River water was expected to dilute the purity of the bubbly stream here, too.

It was like pulling teeth now, getting Susan to tell me anything. During the May-June rainy season, the stream could be thirty to seventy metres wide. And during the July-August flood season, there would be landslides blocking the road, she growled.

We stopped at the town of Xinshan (population 20,000), forty kilometres from the great river. I could not help noticing the freshly built eight-storey building across the street from the bus station. Firecrackers announced the achievement to one and all. I marvelled that here, too, new buildings were rising, even though Xinshan would be inundated. Five hundred families had already moved to the new site. But not everyone would move for at least ten years, said Susan. Still, it did not make much sense.

Susan learned from a bus girl that Wang Zhaojun's home would be spared, but since it was only a "ten-minute" bus ride away, we decided to visit it anyway and then go back to Yichang. The cultural relics bureau was a km away, too far. We shall forget about it, she decided, much to my chagrin. The bus girl said she was happy to move to a new town because the new site was flatter and people did not have to climb so much. Wang Zhaojun's home turned out to be a steep hundred-metre climb through an orange orchard above the bus stop. Her "home" was garden style, very pretty and still under construction. Though she lived two thousand years ago, the first memorial was originally built in the Ming dynasty. It had since been destroyed and the government started rebuilding it in 1979. In the 1970s, archaeologists found two-thousand-year-old bricks and stone animals, since moved elsewhere. We met the director of the site, who said there were now ten locations with ancient relics in the area, but only two or three would be flooded.

Wang Zhaojun, an imperial concubine, is considered one of the four most beautiful women in Chinese history, my friend Yue Chi told me later in Toronto. In pictures, she is usually dressed in furs. Her husband, Kaotsu, the Western Han emperor, had the habit of choosing his nightly companion from paintings of his many wives, or so the story goes. Consequently the women bribed the court painter to make them look beautiful. Wang Zhaojun refused to bribe him and so he painted her to look ugly. The emperor never chose her.

At that time, the Xiongnu nomads, a Turkish-speaking "Mongolian" group (who some Western scholars believe are related to the Kets of Siberia) were attacking their Han neighbours, especially their precious trade caravans. By the third century BC, the Xiongnu had established their own tribal federation, and controlled an immense area from western Manchuria to the Pamir mountains in today's Tadzhikistan. Their leader considered himself an emperor, and even called himself the Son of Heaven.

The Xiongnu were highly successful because of their speedy horses. They were later known in the West as the Huns who sacked Europe. The Han emperors at first tried to control these tribes by force. But then Emperor Kaotsu tried another tactic. He decided to treat them as equals, and to send a Chinese princess as a gift to the leader Modok (Mao-tun).

It is difficult to calculate what is history and what is myth. One hears that when Kaotsu was looking for a princess, he thought he would get rid of the least attractive of his harem, Zhaojun. He saw Wang Zhaojun in the flesh for the first time on the day of the presentation. She was beautiful but it was too late. He had to give her to the Xiongnu. Modok was happy, and the peace was kept for sixty years. But the emperor killed the painter.

Susan told me that Wang Zhaojun had volunteered to go because she wanted to serve her country, but Yue Chi said this was a recent Communist version, written by Guo Moruo to support Communist ideals. Her tomb in Inner Mongolia is a tourist attraction.

The director of Wang Zhaojun's home said 68 families related to Wang Zhaojun still lived in the neighbourhood. We looked for the spring where she washed her face. It still had water in it and I could not resist.

Susan had said getting back to Xinshan would be easy because there would be lots of buses, but we stood beside the road near a small store for a half hour and nothing came or stopped. "It's lunchtime." she wailed. "No one is travelling."

But they were. I hailed the first vehicle that came along, a jeep full of "armed police," a paramilitary arm of the army set up to maintain public order. They had been supervising the construction of small dams in the hills above Wang Zhaojun's home. They do not fight, explained Susan.

The head policeman looked amazingly like a younger version of my own father. Liu Chengshou left two of his men on the road to be picked up by a second jeep, and took us to town, refusing my offered cigarettes. Susan explained that they were there to help people and told them what we were doing. They reminded me of the old image of the People's Liberation Army, the Lone Rangers of China, who went about doing good and then galloping off into the sunset before anyone could thank them. They did not even want me to take a photo of them as a souvenir.

As if fate were on my side, Liu and his deputy, Shi, said they knew the cultural relics people. Their office was near the hotel and the police urged us to stay with them for the night. They would arrange a meeting for us. Their sixty-room building had been opened twenty days before to house local civilian militia during training sessions. It was designed to make money as a hotel at other times. Most government units were required to be as self-sufficient economically as possible, a policy other countries should think about emulating.

The Xinshan Xianrenwubu Hotel was the best of all our hotels. It even had hot water, but only in the evening. It gave away toothbrushes and had big, thick Chinese towels, like the big city hotels. They charged us ¥50 (US$10) for the room, our cheapest and best of the trip. Mr. Liu said that after the Three Gorges Dam was built, the water would go up to the fifth floor of the hotel. They invited us to the most elaborate lunch of the whole trip, and we watched as bottles of the local Wang Zhaojun wine were passed around and the party started to get a little rowdy. They were downright charming. Our host, the manager of the hotel, then told us to leave because they "had a meeting and work to do." I hoped they enjoyed themselves.

Susan slept for the rest of the afternoon, complaining that she was getting tired of me (or at least as she corrected herself), of answering my questions for five days now — a long time! She refused to ask anyone what would happen to the buildings after the flood, insisting that the government would let people live in the unflooded parts. But I learned later that the buildings would be dynamited so ships would not be endangered. I was getting very annoyed with her. I would have fired her earlier but I had only met two people on the Yangzi trip who spoke English. One was Hsu, on the ferry, and the other was one of the managers of the travel agency in Zigui. Susan's English was very good. She had gotten me my bare essentials, though I had hoped for more. Most of my many other guides in China had cheerfully and professionally given service for weeks on end. Most kept me informed of plans and gone the extra mile without complaint. None of them had hinted blatantly about getting a tip. But, I reasoned, she was better than nothing, and she was sick.

Mr. Zhang Daijiu, Director of the Cultural Relics Bureau, and his assistant showed up late in the afternoon, and explained about the relics in the area. There was one nearby, they said, and as we prepared to look at it, Susan said that the morning's walk up the hill had broken her shoe and she did not want to walk any more. Her shoes cost too much. The agency gave no clothing allowance. Mr. Zhang seemed to panic. "It is only two blocks away," he said and she consented to go to translate.

We passed the seven-storey building of the Cultural Relics Bureau. It too looked new. Mr. Zhang explained that the bureau had built it so that when the time came to move, the government would give them lots of money in compensation. So that was it! The government would pay for the building! I admired the foresight, the survival skills. No one need feel sorry for these people! The building was big because it housed both offices and apartments for the staff of eleven. It would be used for the next dozen or so years, until 2007. There was, alas, only one archaeologist on staff.

The relic we saw was only the bottom part of a building and I wondered if it was worth the money to save it. A relatively new house now inhabited by a family had been built on top of it. True, the foundation was Tang. But was this another way to make money, to support the overly large staff of the Cultural Relics Bureau?

Somehow, we started discussing relics in general. Both Mr. Zhang and Susan insisted that relics older than 1795 AD could not be taken out of China. They sounded absolutely certain until I told them about the sale of the five hundred relics at the Shanghai Centre, the oldest piece being two thousand years old. The sale had been co-sponsored by the Cultural Relics Bureau. I also knew of a store in Beijing which sells these ancient artifacts. Susan and Mr. Zhang immediately changed their line and said duplicates could be sold. If I had not known, I would have gone away misinformed.

That night, over my objections, the bureau treated us to dinner.

Next morning, we picked up some buns on the street for breakfast as the hotel's restaurant was not yet open. Susan convinced the police to not only arrange for our bus tickets, but to drive us to the bus station. She flattered the driver, "Why, you look younger than 37!" she said, and he grinned.

I learned later that armed police also provide motorcycle escort for visiting dignitaries, and patrol streets at night in some cities. Later, my Hong Kong-based nephew told me a group in Guangdong had actually battled with the Hong Kong police over a shipful of contraband goods. The Chinese police wanted the goods for themselves. Times are changing.

Sandouping: The Dam Site

(*Robert, Sandouping, September 24, 1992*) As the afternoon light begins to fade the steepest mountains give way to a more open valley. We emerge from the Xiling Gorge as we have from the previous defiles. But where, I ask, is the dam to be built? When we finally arrive at the place it is somewhat of an anticlimax. We have travelled all this way specifically because of the proposed Three Gorges Dam project and we have seen such spectacular sights along the way. I suppose I expect the actual place where the dam would be built to measure up to, and surpass, all the other scenery. But the dam site itself is not, as one might first have expected, in the steepest and narrowest place in the Three Gorges. Rather it is in one of the most open and broadest parts of the river, where the surrounding hills are little more than a couple of hundred metres above the river. I discover later that the reason for this has to do with the nature of the bedrock, among other things.

There are already high-rise buildings standing along the river near the village of Sandouping. These, we are told, will house the workers while the dam is under construction. Now they are empty. There is also an interpretive centre on Zhongbaodao Island where extensive archaeological excavations have been undertaken on neolithic sites. I take a series of wide-angle pictures in an arc across the stern of the ship as we pass the place. That is the only way to capture it on film. But I don't feel that I have captured anything. I have no feeling, no real comprehension, no sense of what is going to happen here.

A couple of hours or so after we pass the dam site we come into the upstream extensions of the city of Yichang. This is the first real city we've seen since Chongqing. I also begin to get some sense of what the lower Yangzi might be like once it escapes the mountains. We have been experiencing the river where it passes through the highlands of Sichuan and Hubei provinces, but after Yichang it breaks out into the vast plain that is home to some four hundred million people. One of the points that proponents of the dam make is that, whatever hardship the reservoir might represent to the people displaced from the upstream valleys, the benefits will be considerable for the vast numbers of people who live downstream.

Yichang will undoubtedly be one of the principal winners, with considerable advantages in industrial expansion and increased shipping. What we can see of the city shows very modern development compared to the other places we've been. It is also built for the most part on a flat expanse of terrain. The river margins here are not wild and rocky, as we have become used, to but are lined with concrete levees. Soon we arrive at the Gezhouba Dam. From the upstream end it is just a long line of concrete, barely rising above the waters of the reservoir impounded behind it. Hydro towers rise above it and power lines stretch off to the horizon.

I worked for several years in the vicinity of the great locks of the Welland Canal, where the shipping of the world is carried around Niagara Falls. The St. Lawrence Seaway turned the Great Lakes into one of the world's most extensive inland waterways, and the history and promotion of the historic Welland Canals was the centre of my work for six years in the 1980s. Because of that interest I have also travelled on and visited many other canals and navigation locks in England, France and the United States, as well as elsewhere in Canada. So the locks at Gezhouba are fascinating to me. There is a set of locks on each side of the dam. The lock chambers themselves are not as large as those on the Welland and St. Lawrence, but the lift, which is accomplished in one stage, is higher than in any single lock in the Great Lakes. I estimate them to be about 180 metres in length, in contrast to the 260 metres of the St. Lawrence Seaway locks. The lift of around 30 metres is just over twice the height that a ship is raised in each of the seven locks of the Welland Canal.

The Gezhouba Dam seems to be a very respectable piece of engineering but, to my eye, there is at least one suspicious aspect. During the time I lived in Niagara one of the lock walls failed. A chunk of concrete weighing several tonnes came loose and plunged to the bottom of the lock. Traffic was disrupted for weeks, and the result was the institution of a complete refurbishing project that has gone on for years and has cost millions of dollars.

The concrete in the Welland Canal locks was fifty years old when the accident occurred. The concrete that forms the walls in the Gezhouba locks is less than ten years old when I am seeing it, but it already looks to me to be in poor shape. Of course surface appearance is not what ultimately matters in engineering terms, nor am I an expert on concrete. But I do think there ought to be some serious concern about the quality of construction methods, materials and workmanship in a project such as the Three Gorges Dam. The existence of the Gezhouba Dam may offer an excellent opportunity to the Chinese to test, monitor and work on improving construction if they proceed with the much larger dam at Sandouping.

The Tujia, the Bo and the Ba

(Ruth, the road to Yichang, November 2, 1992) The public bus trip back to Yichang cost ¥10.20 and took about six hours, stopping every few minutes to pick up or let off passengers. The seats were comfortable, but not as soft as buses at home. We would have arrived even later if we had taken the afternoon boat, and we would have missed the Tujia people.

The trip was through the mountains, past Tujia houses with whitewashed, mud or red brick walls, some of them with unusual brackets supporting the roof. I read later that rich Tujia have wood and tile-roofed houses with paint on them, while the houses of the poor are bamboo and thatch. The Tujia houses we saw were stone with corrugated tin roofs, mud brick covered with mud plaster and whitewash, and tiled roofs (no thatch).

The middle rooms had the tablets of the ancestors and the God of the Earth. Villages used to have ancestral temples but we did not see any. The Tujia are said to be superstitous, believing in ghosts and gods. The towns were small, the road paved. Tujia people are one of China's 55 national minorities, much as the Mohawk nation and the Cree are distinct groups in North America. They settled in the area two thousand years ago. There are several theories as to their origin:

1) they arrived from Guizhou Province to the southwest, which explains why many understand the Yi language;

2) Some scholars say they are the descendants of the Ba people because some surnames sound alike and both were obsessed with tigers, especially the White Tiger God. Both had the same popular family names: Peng, Xiang and Tian;

3) Others say they came as peddlers of handicrafts from Jiangxi province at the end of the Tang and early Five Dynasties (around 910 AD).

Whatever the case, they call themselves *"Bizika"* meaning "local people," or "natives." They call Han people *"Kejia"* meaning "guests." At first the Han called them *"Wulingman"* or *"Wulingxi"* or barbarians, and since the Song Dynasty, they called them Tuding, Turen, Tumin or Tuman.

The Tujia farmed without fertilizers and irrigation, moving when the soil was depleted. Then they started trading with the Han. In 940 AD the headman of the Tujia, Peng Shizhou, sent a six-sided 2,500kg bronze post to mark his alliance with the king of the State of Chu in Xizhou, today's Yongshun County. It was one metre in circumference and four metres high, an impressive achievement.

In the 12th century, the Han people started moving into Tujia territory as labourers, and then as land-owning farmers. From the Han people, the Tujia learned sustainable agriculture, smelting, better tools, and handicrafts. By the end of the Ming and early Qing dynasties, they were using metal ploughshares, sickles and rakes. They knew how to make and use water wheels, a water-powered trip hammer, and a water-powered grinder.

During the Ming Dynasty, Tujia men were conscripted to the southeast coast to fight with the Han, and other minorities, against Japanese pirates. In 1733, the Tujia in Hefeng county rose against the cruelty of their own headman. During 1795-1797, the Tujia supported the uprising of the Miao people against the Han. From 1851, the Tujia set up secret societies and some joined the Taiping uprising. In October, 1956 they were recognized as a distinct nationality, the definition of Tujia changing again before the 1982 census.

Most of China's 2.8 million Tujia people live in western Hunan province south of the Yangzi, where a higher proportion seem to have maintained their old distinct culture. They make their living in agriculture (rice, corn, potatoes, wheat, sugar beets, hemp, tung-oil, and especially tea). They also work in forestry, animal husbandry, wild medicinal herbs, and fishing. They harvest the endangered giant salamander. In Xiangxi, Hunan, they have an autonomous prefecture with the Miao where Tujia traditions are followed. In 1980, two Tujia Autonomous Counties were established in Enshi Preferecture, Hubei province.

Most Tujia now speak Chinese but about thirty thousand (in 1978) spoke only Tujia in the remote areas of Yongshun and Longshan counties of western Hunan. North of the Yangzi, however, aside from the occasional turban, we did not see anyone wearing traditional dress. They seem to have been generally assimilated into the main Han culture here and Mr. Tian, in his rumpled navy blue western business suit and tie, was an example. He understood a little but not a lot of his ancestral tongue.

The traditional Tujia language belongs to the Tibeto-Burman group. The Tujia costume is usually dark blue cotton or sack cloth (*xibu*), the women wearing a short jacket fastened at the left side. They have short, wide sleeves. They used to wear long skirts made from six widths of cloth, but now they wear trousers. The edges of the clothing are decorated with coloured tapes in floral, animal, bird, and geometric designs. Cross (*taohua*) and satin stitches are favoured. They also wear hats or turbans, silver necklaces, bracelets, earrings and anklets. Young men wear baggy trousers and a short jacket fastened down the middle, while older men wear jackets fastened at the side.

Like many other groups, they court now with antiphonal singing. Long ago, a woman could only be married to a blood relative such as her brother's son. A man would be married to his brother's wife if the brother died. Many Tujia people compose songs which express love, or describe battles, work, and bitter lives. There is also traditional dance, the most famous being the *Baishou Wu*, with more than seventy hand movements expressing hunting, fighting, agriculture, and banqueting, which is still performed. Epic dances tell of hopes for a good life and happiness; others of Tujia history.

In the old days, too, the sick would be treated by a shaman, who tried to keep ghosts away and made deals with the gods. The dead were cremated. A shaman would officiate at funerals and kill an ox. Those influenced by Han Chinese later buried their dead, but after the Communist takeover, they cremated again.

There were many taboos, too: girls and pregnant women could not sit on the thresholds of houses, nor carry hoes, wear straw rain capes, or enter a room with an empty water bucket. The sound of a cat should not interfere with the sacrifices to the gods, and cats were not allowed in the same room as the dead.

Fei Hsiao Teng, in his book *Towards a People's Anthropology*, says that the upper class Tujia were used by reactionary Han rulers to dominate the other minorities. After Liberation, the dominated groups refused to recognize the Tujia as minorities.

The 1982 Chinese census showed 2,836,814 Tujia. Of these, 844,435 were considered illiterate in Chinese, and 3,488 had graduated from university.

I learned about the Bo people after I returned home. Scholars think the Bo lived in the Yangzi valley between the 12th and 16th centuries.

There are 22 cliff burial places along the Yangzi, even beyond Chongqing. The Bo people placed their dead in coffins held suspended by the sides of the cliffs 300m up, where no human or wild animal could reach them. Some scholars think the ancestors of the Tujia and Bo people were Ba.

From the many pictographs they left behind, we know that the Ba rode horses, and were farmers. They might have been related to the Yi people of Yunnan province. They were mentioned by Marco Polo in the 13th century. Until the 1900s, the Han majority dismissed these indigenous peoples as barbarians. As a result little has been recorded of their history and culture. Studies of 16 coffins made in 1946 and 1974 revealed two classes of people. The wealthy were wrapped in up to 29 layers of embroidered silk or hemp.

Somehow scholars decided they were contrary people, wearing heavy clothes and warming themselves in the heat of summer by huddling close to fires; and dressing sparcely and holding big fans in the cold of winter. From the skeletons, scholars decided they were of Mongolian origin with Asiatic characteristics. The adults had their upper and lower side teeth removed, a practice followed in this century by isolated tribes in China, Taiwan and southwest Asia.

From legends, we suspect they placed their dead at first in the rivers and later, when convinced by a scholar that they would become more prosperous, started the unusual practice of hanging their heavy five-sided teak coffins.

Later, after doing some more reading, I concluded that the square holes we saw were where the wooden brackets were placed perpendicular to the cliff to support the coffins.

There were so many mysteries to solve, so much to research.

11

THE BOAT HOME

Yichang: The End and the Beginning

(Ruth, Yichang, November 3, 1992) Yichang is at the eastern end of the Three Gorges, a city like most of the communities lining the river, of plain box-shaped buildings clinging to steep hillsides, grayed by coal dust. They were propped against landslides by pilings, against their natural tendency to fall down. They were separated by a few long staircases, or narrow-gauge cable car tracks reaching up from the river, over barren ground that would be covered in high water seasons. None of the towns looked like Tamshui in Taiwan, where I lived in the early 1960s, and which was used as the movie set for *The Sand Pebbles*. None of the buildings here had any trace of lovely porcelain trim on bannisters, or fences, or curved traditional roofs.

Had remnants of the past all been destroyed in the Japanese war? Had the buildings been reconstructed in the cheapest way possible? Were people too poor to put money into decorations? Did they care about the past, about beauty?

The entrance to the Xiling Gorge is where the statue of Zhang Fei, is and the cave of the poets. Susan said that Yichang was the end and the beginning of the hazardous voyage. A pagoda celebrated the safe arrival of travellers there; Yichang was always a trans-shipment point. "Whoever controlled this pass, controlled Sichuan to the west," Susan said.

In 1876, Yichang was opened to foreigners by the Treaty of Cheefoo, which allowed Christian churches, foreign houses, factories and wharves to be built there. She also spoke of the destruction of Yichang by Japanese planes in the 1940s. The Japanese never controlled the Gorges, never fought their way to the capital of Nationalist China in Chongqing. They did, however, almost bomb Chongqing flat. Susan pointed out beautiful, clear Xialouxi Stream, emptying into the murky water of the Yangzi. There was a sharp, diagonal line where the green and brown met. But as the stream was below the new dam, it would not be affected.

We visited the multipurpose Gezhouba Dam in the eastern part of Yichang. It had been opened in 1988 when the population of Yichang was 100,000. Now, four years later, it is 470,000. The dam produces 14.1 billion kwh of electricity annually.

In Yichang it took about 12 hours to telephone my husband in Canada from the hotel. Apparently, there were only a few long distance international lines from the city, and during the day all were busy. I finally managed about midnight. Even though it was the most modern city between Wuhan and Chongqing, Yichang obviously needs lots of work and help.

Ticket Trials

(Caroline, Yichang, September 24, 1992) We are off the boat at Yichang's new terminal by 4:30. We need an action plan. We have counted and re-counted our remaining FECs and RMB. One idea is to split up: half go to find the ticket window, half to the bank.

The terminal has everything Chongqing has only dreamed of: a waiting room, washrooms, stores and ticket offices. Which one? There is a chalk board, with what look like ship schedules, but in Chinese. Then there are at least four wickets with long lines. Which one? Our friends hang back. In their experience, this business is always done by a designated person in their work unit. If they were buying a ticket for themselves, their usual way would be to seek out a friend, or a friend of a friend who works here, a schoolmate or an acquaintance of a schoolmate. This strategy is based on the old notion that tickets are so rare that the only way to get one is through special connections.

Kailin is especially reticent. Our argument over laundry has evolved into a discussion of why men buy tickets and women order lunch. Both women have decided that buying tickets is Zhongling's job, and they sit down with the pile of luggage. So it is my job to get Zhongling organized, to get him to approach the ticket counters. Zhongling is reluctant to stand in line and wait his turn. First I suggest that each one of us go and stand in a line. When one of us gets the front, a Chinese speaker will step in and ask the clerk if this is the right place. This idea is a non-starter, as is consulting with those already standing in the lines.

Zhongling will play the "foreign card." He learns from talking to someone that there is a foreigner's office upstairs. All of this is frustrating because time is passing (the last tickets might be sold while we stand here arguing, and the bank for sure is closing). I attach myself to Zhongling and follow him to the very end of the large hall and up the stairs to the second floor. Is this CITS? Even that would be fine at this point. There are human beings we can talk to, if we can get their attention, that is. Better than shouting over an intercom to a person behind bars and windows in a ticket wicket, I concede, if it works.

Finally someone comes. Zhongling translates. There is a boat tomorrow night with second-class vacancies. I say, no good. How about tonight, in third class? I suspect that Zhongling has only asked about tomorrow, wanting to stay in those ¥200 hotel rooms. Yes, there are vacancies in second class, on a boat boarding at 10:00 p.m., leaving at 2:00 a.m. tomorrow morning. The price is FEC ¥390 for foreigners, ¥186 for Chinese. By borrowing a bit from Zhongling, we will have enough cash. Not that the tickets can be bought here. We must go to wicket number one. Not surprisingly, the wicket for second (first) class has the shortest line. In ten minutes or so, it is done. The negative part is that we will be three nights, arriving at noon on Sunday. But compared to the alternatives, I have to admit, that is fast. And easy and cheap.

The voyage from Yichang to Chongqing costs less than the reverse: and is considered less interesting. Very few make the round trip; they start from Chongqing and go on from Yichang or Wuhan. Building the dam will bring modern airports, especially to Yichang, and that will make travel for business much quicker and will increase tourism on the Gorges. Now tourists spend a day, at least, getting from Yichang to Wuhan and its airport.

What next? Find "Left Luggage," see the town, have supper, get on board. There is a left luggage, of course. We are about to set out on our adventure when Zhongling remembers that he has checked his face-washing kit and we must wait while he retrieves it. Face washing sounds like a good idea. The water in the terminal washrooms is shut off and the sinks are full of garbage, but there is a tap at the outside far corner of the terminal. So what else is new? One by one we sponge and splash our sweaty arms and faces under the tap. Then we all watch with fascination as Zhongling washes his face with soap, all over, dries it carefully with his towel, folds the towel carefully and puts everything away.

Next is food. If we borrow some of everybody's remaining cash, we won't have to spend time trying to cash traveler's cheques. The bank is long since closed, but one of the ¥200 hotels may have a foreign exchange desk (it may not, however, cash cheques for someone who is not a guest). We cannot immediately find a phone, no pay phones, and we do not know where any of these hotels are. (Again, why didn't we just ask?) Zhongling is ready to negotiate, but we say no-thank-you to a throng of bicycle rickshaw drivers. We have been on ship all day, and will be for three more days, so we need a walk. (In fact, we are doing Zhongling an injustice. A bicycle rickshaw would have been able to take me to one of the big hotels, but we are not thinking much.)

Yichang is the biggest place we have seen since Chongqing and it seems like Paris to us. The streets and sidewalks are wide and lined with trees. Crossing the streets involves avoiding the bicycle riders, riding in self-absorbed slow motion. There are sidewalk stalls everywhere, making

roasted chestnuts, noodles, and flatbread. Li and Kailin exclaim on the
quality and variety. Prices are lower than Chongqing. The clerks in the
packaged food store wear white lab coats and hats! They want to shop.

Then, we see a restaurant advertising "local specialties." It is a bit of
a lie, but the proprietor is another compatriot from Chongqing. Not a recommendation to my mind. I am beginning to think that these men are
smalltime promoters who do not give good value. Robert is feeling a bit
rocky, the hot pot has done his digestive system in too, and suddenly no one
wants to walk any farther. Although it is early for supper, and we are the
first customers, we pile in and make ourselves at home on the tiny wooden
stools. Kailin and I do our best with the menu, looking for dishes that won't
aggravate our stomach troubles. Sichuan-style Fried Fish comes as boiled
catfish, gray and soggy, obviously a mistake, and we send it back. Eggplant
with fish sauce arrives with soya sauce, but this is possibly a misunderstanding between Kailin and me and so we keep it. Sichuan Green Beans is
pork and canned peas; sweet and sour chicken arrives as stewed chicken
with gravy, but it is good. The fish dish eventually reappears, and more or
less redeems the cook.

The food is, at best, just so-so. We are in a state of physical exhaustion
and general emotional letdown, and the fact that our host takes our order
and then serves any old thing does not help. He is also using "just in time"
inventory management practices: each drink we order has to be specially
fetched from one nearby establishment or another. We order Gezhouba
beer, to try the local brew and toast the Gezhouba Dam.

Our host is eager to chat. We have asked for the no smoking section,
but he pulls up a stool, cigarette in hand. Yichang has already had a boom
during the building of the Gezhouba Dam and is looking forward to a much
bigger one this time. It is the big city nearest the site. Too bad for
Chongqing; the goods and services will move through here. He does not
know how many people will be resettled here, but Yichang will grow. We
should buy some real estate!

The night is cool and pleasant and Robert and I are eager for some
exercise. Our friends are eager to buy presents for their families. Both
Kailin and Li have young sons, and their husbands and mothers have been
looking after things in their absence. The first block features noodle and hot
pot restaurants. Their dishes look tastier and fresher than what we have
just consumed. The next corner, however, is the night free market. Mostly,
there are clothes for sale, dark polyesters. Robert is still looking for a child's
replica of a Red Army uniform for his eight-year-old son, Ceilidh. There
were lots of them in Shibaozhai, but not when there was time to shop.
Kailin tries hard, but remarks that they must have been last year's fad and
so remain only in the small towns. Finally, we find one that may fit.

We also find a child's black satin mandarin hat with a woolen pigtail,
like the one I had bought up river. This one is part of a suit. Dragons are

embroidered on the front of the little jacket, and the pant legs are decorated with Mickey Mouse! I want it. No one will go below ¥35. Aside from carrying it, we are low on money. The vendor finally comes down to ¥25, but regretfully I pass. Elsewhere, I do buy another pigtail hat and it is a surprise hit with a friend's year-old baby back in Chongqing.

The next attraction is a blanket spread with copies of Tang pottery camels, horses, Du Fu on a Water Buffalo, and a marvelously anachronistic *Guanyin*, about half a meter tall, for only ¥38. Never have I seen such a *Guanyin*. I try to convince Robert that it would be a perfect gift for his wife, Pam. Where would they put it? Make it into a lamp! Wrong kind of Buddhism.

Zhongling and Kailin like the Tang figures too. A woman with a computer, in Chinese and English, says she can use it and an electronic sensor gizmo, to diagnose our ills. There are pinups from Hong Kong, odds and ends of clothing overruns from joint venture factories on the Coast, sunglasses .... People have set out bowls of water under the light bulbs to catch the fruit flies that are everywhere. Here, for the first time in over a week, we see foreigners, just off the cruise boats. They look tall and pale and blotchy and fat. A good thing we cannot see ourselves!

Robert is not much interested in the sociological implications of the goods for sale in the free market. I duck into a tape store. The new Michael Jackson? No, classical Chinese, classical Western, and the equivalent of our Broadway musicals that Kailin likes to sing, including *The White Haired Girl*. In fact, both of us have almost had it, and I tell the others that we will meet them in the waiting room at 8:30, reminding Kailin to buy a lot of fruit to take on board.

The waiting room has a no-smoking section (most of its occupants are women). The lights blink up and down as the power ebbs and flows. Gradually over the day, the grit and garbage have piled up. Someone, somewhere in this great centralized country, designed a wonderful set of litter receptacles. Gaily-painted cast iron frogs, pandas, what have you, sit astride bins that hold at most a few liters. They are everywhere and they fill up in about half an hour in the morning. For a little while people pile their garbage beside them. Then they give up and drop it anywhere.

A man sits down beside me and lights up. I say, *"Bu chou yao,"* in my best Chinese. He moves. A smartly-dressed and expensively suitcased group gathers. Will they be on board with us? We watch them disappear into the first-class waiting room. Every so often a bell rings and people rush out the far doors. There are four docks below the terminal, but the boarding process is the traditional one. We are too tired to write and it is almost too dim to read. What keeps us going is the thought of a nice clean bed.

8:30 p.m. No one has come back. Robert goes to look. No one nearby. 8:45, I go. Each time we must run the gauntlet of the rickshaw entrepreneurs. Our group is sitting in a store on the nearby corner, in a great sea of plastic bags. Li has bought an immense melon, but only a few oranges and no bananas. Kailin has carefully appraised every pear and apple in town and they have all been found wanting. They do have peanuts, Sichuan pickle, salty duck eggs, several pairs of slippers (¥ 4!) and many, many other presents. How are we to carry all this and our baggage too? My dismay does not dampen their triumph. We gather up all the bags and make our way back through the rickshaws.

It is 8:55. We must call Norman to tell him we are on our way back. I have located the telephone centre, a couple of buildings over. By the time we arrive, however, it is 9:04 and the staff has stopped making calls. The old, sick foreigner plea falls on deaf ears. Well, Zhongling says, we can get a taxi to one of the hotels. That means finding a taxi, finding the hotel, getting a call through, and taking another taxi back to the dock to board at 10:00. Zhongling had promised to be back at 8:30! Group discipline says there is not enough time, and after the shopping spree there probably is not enough money. I am bitterly disappointed.

By 9:30, I suggest we get our luggage: all we need is for "left luggage" to close. In fact, it is open all night. The boat, of course, is late. The washroom is unlit and the gutter where you pee is piled with used toilet paper and sanitary napkins. The water is off and there is no water to wash the shit away! Cigarette butts, garbage, everything is piling up. The couple of peasants with twig brooms and bamboo baskets, whose job it is to clear the trash, whisk away halfheartedly.

I eat one of the oranges and make a point of depositing the peels in one of the garbage baskets. Then I slump down in my chair. I have been awake since 4:30 a.m. and it has been a very hectic week. I am brought back to consciousness by a tug on my clothing. On the floor beside me is a little creature, with spindly chick legs, a ball of a body, two tiny arms and big sad eyes, holding a one *jou* note for me to see. Shuffle, shuffle. He finds my eyes. I look at Robert, dozing. He says, "Ignore him." Kailin is implacable. I avert my eyes. He finally shuffles off.

Paying Lawyers or Paying for Wildlife Biologists

(Ruth, Yichang, November 4, 1992) My travelling companions on the train from Beijing had mentioned the effect of the new dam on fish. The pink river dolphin (*baiji, Lipotes vexillifer*) has been labelled the "most endangered dolphin in the world" by a recent conference of the International Association for Aquatic Animals' Medicine. There are said to be only 150-200 of them left in the wild, in the Yangzi, too small a gene pool to revive the species. Only two dolphins were spotted during a recent 660-hour, 4,700km dolphin search by a university team.

Qi Qi, at Wuhan's East Lake, is the only one in captivity. These 2.3m, blue-gray, long-snouted dolphins live near Wuhan, the provincial capital downstream of the dam. There they are further threatened more by Wuhan's yearly 12.3% increase in industrial output and the Changjiang (Yangzi) Economic Zone. Some are killed by trolling fishermen, nets, electric prods, explosives, or, like Florida's manatees, by boat propellers. The sand bars where the dolphins rest and tend their young would be damaged. Will the new dam be the fatal blow to an animal many Chinese used to consider supernatural?

The dam itself *might* affect the migratory fish on which the dolphin feeds and harm other food supplies and resting areas of the dolphin and the Chinese alligator *(Alligator somemsos)*. Higher water levels might prevent the alligator from burrowing into muddy riverbanks. These two-metre-long creatures are found in the lower Yangzi Valley, and are also almost extinct.

The 62cm-tall pink-faced, golden-eyed, Siberian White Crane (*Grus leucogeranus*) feeds on a weed in the bottom of the river downstream. With changes in the water level, its food *could* be destroyed. Other endangered species listed as in the dam site area are Temminick's cat (*Felix temminicki*), the stump-tailed macaque (*Macaca arctoides*), the Sichuan snub-nosed monkey, the Chinese giant salamander, the Cloud Leopard and the siku deer (*Cervus nippon kopschi*).

I learned about these animals from Defenders of Wildlife, a Washington, D.C.-based organization which has been suing the U.S. Interior Department. That department did not consult with the Fish and Wildlife Service before agreeing to allow the U.S. Bureau of Relamation and the Army Corps of Engineers from entering into a five-year extension in December, 1991 of their contract to help design and build the Three Gorges Dam. Such a condition, they say, was a mandate of the American Endangered Species Act which ordered consultations "on any action that may affect an endangered or threatened species." An amendment in 1986 exempted projects overseas, but the Defenders of Wildlife claim the amendment is illegal.

The Defenders seem primarily to be lobbyists, relying on other organizations to do their research for them. Judging by the information they sent us, their scientists need to spend more time on the Yangzi. They say that rising water levels will affect the Giant Panda in its mountain habitat, but this is hard to conceive. One wonders if they are exaggerating the effects of the dam, since it is upstream of most of these endangered species. If they are genuinely interested in saving wildlife, perhaps they should be spending their money financing breeding facilities, rather than paying lawyers.

Something is being done about the pink river dolphin.

They did not mention that a 2,666 hectare, 21km-long semi-natural reserve is being set up in Hunan Province (at Tianezhou), hopefully for 20 to 25 dolphins. Another smaller such reserve is at Tongling in the lower reaches of the river. A US$2 million breeding facility is also being built. Only the future can decide how effective these will be. I have heard of one alligator-breeding facility that is so poorly funded that the alligators are starving to death.

The International Crane Foundation seemed to know more about its subject. Its director George Archibald wrote: "With the exception of ten Siberian Cranes that winter in Iran and five that winter in India, the lion's share of the species (about 2,500) spend the winter on Dongting and Poyang Lakes along the middle reaches of the Yangzi. . . . Of the world's 15 species of cranes, the Siberian Crane's survival is most closely linked to the avail-ability of vast expanses of shallow water. . . . The effect of the Gezhouba Dam on populations of waterfowl downstream remains unknown. How-ever, there are concerns that the proposed Three Gorges Dam will retard downstream summer flooding to such an extent that these two lakes will not fill to levels that support the growth of wild celery beds upon which tubers the cranes forage in winter. In addition, siltation deposited behind the dam will mean less silt deposited on new land along the coast of Jiangsu province, essential for the winter feeding grounds of the endangered Red-crowned Cranes."

"Our Chinese colleagues tow the party line that tributary rivers will provide adequate summer flooding in the two great lakes. However, it is the swell of the Yangzi that eventually overflows into the lakes that completes the process of seasonal flooding. If the flow of the Yangzi is reduced, the level of flooding may be greatly reduced."

We do not know if it will be reduced.

Chinese and American scientists have both been wrong before. Cap-tive breeding might be the only measure to ensure the future of any of these species, and the Chinese, with outside help like Hong Kong's Ocean Park, are working in this direction, at least with the dolphins.

In Yichang, I visited the Chinese Sturgeon Research Institute, which raises the unique Yangzi River sturgeon and releases them by the millions into the river east and downstream of the dam. The sturgeons (*Acipensor shenris*) are huge, ugly and knobby, with long pointed noses.

The Yangzi River sturgeon is not mentioned in the litigation by the Defenders of Wildlife, as it is not on the U.S. list. These sturgeon date from the time of the dinosaurs. They are unique to the Yangzi and it is forbidden to kill them. We saw some live ones about two metres long and I felt reas-sured that the government was concerned about these fish, although how many actually survive after release into the Yangzi River remains a crucial question.

Patient Tired Eyes

(Caroline, Boat back to Chongqing, September 24-26, 1992) Time seems to have stopped. At 11:30 the ship arrives. We consolidate the shopping bags

and put as much as we can into our luggage. Our dock is the furthest, but the anticipation of sleep provides one last shot of adrenalin. We arrive before the other passengers have gotten off and so we must put down our bundles and wait. The fifth-class passengers are in line ahead of us, and they must have time to settle with their immense bundles. Most of the wide places in the stairways and halls are already taken, their owners asleep on top of their sacks and boxes.

Finally, we are let past. Up four flights of stairs we go, to the fourth deck, stepping around and over, and we hope not on, these people. The stairwell is in the middle of the ship, above the engine room, and the fumes and the heat seem unbearable. And still, each one clings to his or her scrap of space. A young mother and a solemn, silent baby, both with patient, tired eyes, stare at the foreigners with their filthy clothes, their cameras, and all those plastic bags.

Each cabin has two beds. Li agrees to share with a stranger. That stranger we label the "concubine" of a Hong Kong businessman, a pretty young woman in pretty clothes. He spends little time in the cabin (I guess that is not allowed) but he is always coming by and Li does not get much sleep. (The next day they get off and after that Li has the cabin to herself.) The cabins are all inside cabins. The area is roofed in and so we can visit each other or the bathrooms without getting wet. Everything is neat and clean, including towels and wash cloths and complimentary soap, tooth brushes and paste. Three nights and two and a half days cost foreigners $100.00 each. Half that for locals.

Kailin can see that I have about had it and helps me get organized and into one of the washrooms. I cannot get the heat and the fumes and those tired eyes out of my mind, but our friends are delighted with their cabins. And so there I am. Cotton shift, soap, towel, and shampoo, all in a bundle, and I am even getting the hang of the slippers. Beside the bathtub is a shower and below it a drain. I finally realize that these bathrooms are designed Japanese style. One takes a shower before getting into the tub! The tub has a plug, but the enamel has long gone and it does not look too clean. The shower is simplest. Is the tap on the left the hot? Well, the one on the right is colder. (Kailin and Li have opted for the "Chinese" washing room, where there is a basin and really hot water.) I don't care. I soap or shampoo all over, stand under the lukewarm drizzle, and then get into the rough red gown.

The argument over the laundry begins again. A Chinese woman near to running water cannot be stopped. Tomorrow, I promise. There is no window. I put on the fan but Kailin does not like it so, soon, I put it off.

Sleep is fitful. The boat shudders. The blasts on its horn are loud and long. Each new volley wakes me up. The engine strains against the river, but surely we cannot have passed the locks and dam. Kailin says she woke up early and turned on her reading lamp. I don't remember that so I must have gotten some sleep. My head aches and I keep seeing those solemn eyes and the little chick-person.

I awake next morning with a headache. Too late for breakfast. Headache pill, orange, banana, ice cream sandwich cookie, hot water for coffee. A quiet day. The second-class section has few other occupants. The lounge has a long wooden table, plastic bonsai trees, and windows that look out on the forward deck and the river. It is a refuge, a wonderful place to write when it is not full of smokers or noisy card players. Kailin loves it. The good weather, now that we do not need it, has vanished. It is dark and gray. I am not ready to face the bodies in the stairways. The others go for food. Lunch is hot green peppers and greasy pork. Robert, who does not like green peppers has "Swedish" pork balls. Supper is pork and huge pieces of ginger. All but Li pick out the ginger.

The argument about laundry recommences. I suggest we wait until we get home: I do not want the cabin hanging with dripping laundry, hitting my face, making the rug wet and muddy. Kailin agrees to hang the laundry outside. Other people are doing the same, and soon there is a small forest of dripping things at the end of each hall. Then I refuse to let her wash my clothes. She says, "Women wash, men buy tickets and carry heavy things." I say, "Feminists write books."

Robert and I suggest she write an article about "the crazy foreigners," the differences between Canadians and Chinese, for a travel magazine she mentioned, and for the book. She had asked us to do an article for the magazine. We refused. Earlier she had said she could not do the article because she was too busy. "Too busy," I said, "doing a foreigner's laundry! Bugging a foreigner about doing her laundry!" She threatens, "We are not friends." Finally she relents and has a lot of fun, consulting with Li and Zhongling about our foibles. Looking up words. I have hardly ever seen a Chinese person write anything without their much-thumbed dictionary of Chinese characters.

After supper, we have a meeting. Li is to tell us about her participation in the dig at Zhongbaodao Island. There are many questions not answered. Is there a database of sites being made? By whom? In China, in the old days at least, the names of people on the committee, if not the committee itself, might have been a secret. I do not think it is like that today, but our friends very often think it is. They have a habit of self-censorship. We hope that Ruth can visit the officials of the National Bureau of Relics when she goes to Beijing in October.

Li is interested in graveyards because that is what she will be excavating later this year. The final thing Li says is that, while she will help us by finding articles, she does not want to take any active part in our book. That is enough meeting. We borrow the tray from the lounge and cut up the huge sweet *hani* melon from Xinjiang Province. Robert goes to the snack bar (finally a boat with one) for a bottle of beer, but no luck.

We have one more adventure: the potsherds in our luggage. Kailin's question, "What are you going to do with them?" looms large. One thing we can do is photograph them. The operation is carried out in greatest secrecy. Robert brings the tripod, the 64ASA film, and what lighting we can arrange into our room. Then, one after another, we take out the shards and the stone tools and photograph them, along with a slip of paper on which we have written the name and period of the site. Li arranges them in little groups of two or three shards from similar pots.

That is not the end of the story. When we are back to living in a Chongqing hotel room, Zhongling comes to speak to Norman. Norman is Chinese, he will understand. Caroline must give back the shards. Li is entitled to have them but none of the rest of us are. They could go to jail. Well, I had yet to get around to figuring out just how to approach the problem. This solves it. I hand over the shards, still carefully packed in their plastic bags. What hurt the most, was Kailin's comment: "They aren't as pretty as Li's." To me, they were.

A Lesson in Relativity

(Robert, on the boat to Chongqing, September 25, 1992) By the time our long slow trip back upriver begins we are already veterans of travel on half a dozen different river boats. We are beginning to get to know them almost too well. We see a couple of the classy tourist ships from a distance but that is as close as we get. They are about the same size as the regular river boats but carry fewer passengers, in considerably more comfort. It isn't until I arrive back in Canada that I will really understand the degree of difference. The local newspaper editor in the next town tells me that he has travelled on one of these luxury boats with about forty other people. The boats we take must at times carry close to four hundred souls.

The regular boats with their four classes, second to fifth, are a study in the persistence of a kind of feudal system in China. Because it is nominally a socialist country there is no first class, second being first. There the lip-service to equality ends. When we board Steamer Number Nine after midnight in Yichang we check into second-class cabins. That completes our tour of the four levels of service, since we have had the opportunity of enjoying each of the others during the previous days.

With two people to a cabin and relatively clear showers and toilets (relatively clean), this boat is the pinnacle of comfort for us. But for some reason the disposition of the other passengers, particularly those in fifth class, is particularly horrific on Steamer Number Nine. Perhaps it just appears this way to us because we are tired and worn out. Some of the peasants travelling on the decks have already been aboard for days when we arrive. They are crammed into the narrow interior passageways because it has begun to rain quite hard. They are sleeping on top of the bags they carried on or on tattered blankets spread on the cold steel of the deck. Whenever we leave the second-class area we have to step over sleeping families and clutches of men playing cards. Caroline doesn't go out there for two whole days. Sometimes the culture shock is just too much. It overtakes you and makes it impossible to face the stark realities of China with a capital 'C'.

The one place where egalitarianism triumphs aboard the Yangzi River boat is in the cafeteria. Once again Steamer Number Nine surpasses all others in this regard. Twice a day on the long run back up the river toward Chongqing we make the trek down two decks, over the blue-clad figures sprawled in the passageways and through the forest of drying laundry to the food line. Zhongling buys a handful of meal tickets at the cashier's window and then we wait to see what fare is on offer. Although I have come to realize that eating is an important part of Yangzi culture and have enjoyed many great meals, there is another side to the equation.

I am reminded of Bill Purvis's comment in his book *Barefoot in the Boardroom*. He talks about food in the factory cafeteria at the Gold Land Foundry in Guangdong Province. He says these meals would cause a riot in any prison in Canada, but are accepted without protest by the average Chinese worker. I suppose one ought to remember that in China millions of people have starved to death, even in this century, and virtually everyone over the age of twenty has probably known real hunger. The availability of any food in quantity must be seen in this light and the acceptance of it by ordinary people should be judged as a tribute to their incredible patience and perseverance. Having said that I can simply comment that the quality of the grub on Steamer Boat Number Nine deteriorates steadily as we plow doggedly back up river.

I don't know whether they are just running out of supplies or what, but the first day there is greasy pork and limp vegetables with ginger. The ubiquitous thin soup of tomato skins and runny egg runs out the second day. On day three the pork disappears. By that night we are down to ginger. Not ginger as a spice or ginger as a garnish, but a meal made up entirely of big, tough, stringy chunks of overly mature ginger root, fried up and plopped on top of our helping of rice. It is too much for Robert. My digestion is already in shambles. I give up on boat food altogether and survive for the next day and a half on the remaining oranges we have with us and a bottle of beer bought over the rail from a riverside vendor during one of our stops.

If the lower decks of a Yangzi River boat represents the down side of the remnant feudal mentality, then the second-class lounge shows not the up side, but the top side. The kinds of "high party officials," as Zhongling characterized them, who are allowed to travel here act like overlords. When Caroline attempts to get Zhongling to ask a rather fat ugly smoker to move from his seat directly in the doorway, Zhongling refuses.

"Are you afraid of him?" she asks.

"Of course I'm afraid of him," Zhongling honestly admits.

This is just one more example of the general deference to authority and privilege that I see as typical in China. Even in our little travelling group there is often a preoccupation with deciding what is the seniority or hierarchy. Often when we ask questions the friend asked will defer answering by saying that they are not experts. One of the things that most puzzles our Chinese friends is our willingness to explore and question and speculate about virtually anything we see or experience. This concept of what I think of as "environmental citizenship" or the right and obligation of any citizen to question and understand what is going on around them, seems as foreign to the Chinese as our high-bridged noses.

"In what area are you actually an expert?" is a question put to me more than once.

"I know about a lot of things to a greater or lesser extent," I answer. "And what I don't know about I try to learn." But what I really am not able to communicate is that I aspire to be an expert "citizen," of my own country and the world. I want to be able to, and to be free to, figure out what the heck is going on in the world around me and what I can do about it. I wondered if, as the Chinese import such Western treasures as consumer advertising and Christmas shopping, will they be allowed to develop, albeit in their own way, this concept of citizen participation?

Helping: Benefits Outweigh the Disadvantages

(Ruth, Toronto, November 7, 1993) If the Three Gorges Dam project were cancelled, there would be no need to save the relics. Those underground would continue to sit there for another thousand years until some enterprising builder or grave robber came along, or until the Chinese government had the funds to excavate them.

But the decision has been made. People have already been moved and construction has already started. It looks as if the dam will be built. The question then becomes, do we want to be involved? If we do nothing, will relics be lost? The Chinese will do their best to save what they can, but they cannot save everything. In the light of other, more pressing problems, do we care if some human knowledge is lost?

There will not be another set of terracotta warriors. The Yangzi Valley was not the burying ground of emperors. But it was the battlefield of the Three Kingdoms, and people tended to bury their valuables to save them from marauding troops. And then there are the graves. What would the grave of a general, or the local mandarin reveal? What would they need to help them in their life after death? More than one magnificent set of bells have already been found.

And will all the money donated be used wisely? Will some of it go to corrupt officials or fancy housing for the Bureau of Relics?

I go to China because I want to help. But I also want to learn. There is much to experience, a different culture, a different way of thinking, a country in flux between stagnant tradition and a new conception of freedom. I want to figure out why the economic growth rate here is so much faster than in Canada. I want to contribute to the development of both countries.

But I think the need is not just saving relics, but building modern museums in which to store the relics. There is no point in digging up treasures and putting them into easily burglarized buildings. Hundreds of thousands of relics are lost every year through poor security, bad lighting, inadequate climate control, and wicked officials. China needs vaults and reliable computerized inventories and staff training. It needs trusted guards around archaeological sites. Over forty thousand ancient graves were robbed in 1989 and 1990.

The government cares very much about stolen relics. Recently they executed two grave robbers and arrested thousands of relic smugglers and looters. They caught the former head of the Cultural Relics Bureau in Xiamen and gave him life imprisonment for stealing 88 relics from the museum and library, for which he was responsible.

But China currently has ten million items in fifteen hundred museums. Most of the museums have inadequate security systems. There is little to keep the farmers who own the orange trees from digging around, looking for relics and selling their finds to unscrupulous dealers and collectors. In 1992, the Chinese government spent ¥120 million protecting and rescuing relics. In 1993, it expects to spend ¥130 million. With the additional burden of the flooded areas, it needs help.

How would I like to do this? I would want to make it possible for classes in schools and archaeological societies around the world to adopt a site and follow the whole archaeological process.

We need money not just for the archaeologists but also to write letters to businesses, to international organizations, to universities and foundations. It takes money to raise money. We need to get volunteers and pay for airfares. We need to coordinate activities with other countries. Saving Yangzi relics is an awesome project.

My experience with Susan, who was one of the worst guides I have had in 27 years of travel in China, is a reminder that however frustrating we may find it, things have to be done keeping in mind the Chinese way. Her hints at how poor she is, how much other people have tipped her, and her reluctance to translate, fortunately are not general. They do, however, indicate the new "me first" generation. The economic and political changes in China have come quickly, and a new morality has not yet evolved to control it. Her's is not the imposed selfless dedication of the Cultural Revolution. Hers is an impatience to get what her parents' generation could not have, what China's openness to the outside world is telling her she can have.

But she has not yet thought through what her selfishness means for her family, her town, her country and the world. Anyone working in China must be astute, aware of these new social and moral problems. One must not be intimidated by undue requests for money or help to emigrate or study abroad. One has to know something about China's background to understand the current metamorphoses, to work with Chinese people, and help them avoid mistakes made elsewhere. There are many dedicated, self-less individuals. I like to think that Susan's heart is good. Not everyone will feel my pull to help China. Each one of us has our own thing to do. But if you think your destiny is in that fantastic country, I can only repeat the old refrain about the dam — the benefits outweigh the disadvantages. And do not worry, transportation, telecommunications, hotels and translators should improve soon.

Think also of the future generations who will be grateful for your help. Think of the scholars who might be able to come to new conclusions about the date of the first smelting of bronze or the first written language. Think of exquisite pieces of sculpture and ancient coins that might be saved. Think of medical techniques used hundreds of years ago that could be used today. Recently we have heard about an ancient Chinese remedy that stops the craving for alcohol. There are the ungerminated seeds for seed banks that might be the answer for world hunger. And what about DNA samples for future gene banks? The list is endless.

But at least you can take a look now at a doomed landscape and see for yourself something that will never ever be seen again in the same way.

Li and Yi: Righteousness in the Face of Profit

(Kailin, On the boat back to Chongqing, September 26, 1992) The most important question for the Three Gorges Project, I think, is not a particular family's opportunity or inconvenience, or the appearance of some "false temples," but the development of China's water resources. Can the harnessing of the Yangzi River solve the fundamental problems and eliminate the scourge of floods? There have been many opinions and the debate goes back many years.

It ended, for now at least, at the National Peoples Congress and Communist Party Central Committee sessions, April, 1992, in the Great Hall of the People in Beijing. Some deputies and members mentioned some problems. As Lei Hengshun, a deputy from the city of Shangqing, said, "You can't find any reservoir in the world with its end point lying beside a major city: the reservoir of the Three Gorges Project will end at Chongqing, one of the most populous areas in China. This will result in some special problems such as heavy deposits of sand in the port and a rise in the water level to a point which can threaten the safety of Chongqing."

Central Committee member He Shaoxun, a professor at the Changnan Industrial University in Hunan Province, said, "Apart from my fear that the project might become a military target in future wars, I was also worried about problems such as avalanches, landslides, earthquakes and population movements." Xu Bowen, a National Congress Deputy, suggests that the State pay full attention to the problem of water and soil conservation on the upper reaches of the Yangzi River and that rampant tree-cutting be prohibited to reduce sand deposits in the reservoir. Qing Changgeng, a deputy from Yunnan Province, also suggested that investment needs to be estimated more accurately and that more scientific work be done.

But others feel it will strengthen our country and make people rich, after the inevitable problems of unexperience, corruption and malfeasance are solved. If these questions can be answered, the project will be beneficial to the whole country. We can turn for some guidance to the fine tradition of Chinese Culture. We can learn from what Confucius (551-479 BC) said in *Lun Yu*, "Consider righteousness in the face of profit. Gentlemen understand righteousness, villains know profit. Be careful if you want to get profit. You can get it if you understand righteousness and if you are not impatient."

Righteousness refers to an appropriate, right, correct, suitable moral conception. Ancient Chinese philosophers advocated a preference for *"yi"* (righteousness) over *"li"* (profit), which today would mean public versus private profit. While righteousness implies moral standards, and these may change with time, the principle of upholding righteousness is always there. In opposition to the individualism advocated by the West, Chinese believe that the value of a human being is realized only in his relationship with his or her fellow beings. This concept integrates benevolence, righteousness, tolerance, harmony, sense of duty, and a contribution to a great consensus or collectivity in which the destiny of the individual is closely related to that of the society in which he or she lives. And it encompasses the sacrificial spirit in their contribution.

Today, the Three Gorges Project needs people to sacrifice their individual and small collective interests, and to serve the interests of their country. We believe that people will not begrudge this. We learned from a farmer outside of Fuling that the sum of money given by the government for relocation is more that the cost of rebuilding the house. The individual and the small collective won't lose money or opportunity. From this we can understand that the burden on the government is heavy and its determination is great.

Bibliography

An Zhimin. "Archaological Research on Neolithic China," *Current Anthropology* Vol. 29, No. 5., 1988 (753-759).

Ballard, W. L. "Aspects of the Linguistic History of Southern China," *Asian Perspectives* XXIV (2), 1981 (163-183).

Bellwood, Peter et al. "New Dates for Prehistoric Asian Rice," *Asian Perspectives*, Vol. 31, No. 2, 1992 (161-170).

Bird, Isabella. *The Yangtze Valley and Beyond.* London: John Murray, 1899, reissued, Virago Press, 1985.

Boyd, Andrew. *Chinese Architecture and Town Planning.* London: Alec Tiranti, 1962.

Brooks, Alison S. and Bernard Wood. "The Chinese side of the story (paleoanthropology)," *Nature* Vol. 344, 1990 (288-89).

Chang Tetzu. "The Ethnobotany of Rice in Island Southeast Asia," *Asian Perspectives* XXVII 1986-7 (29-34).

Cooper, Arthur. *Li Po and Tu Fu.* Harmonsworth UK: Penguin, 1973.

CPAM, Sichuan Archaeological Institute of Sichuan Cultural Bureau et al. "Excavation of the Sacrificial Pit No. 2 at Sanxingdui Site in Guanghan," *Wenwu* 1989(5) (1-20). In Chinese.

Crespigny, R. *Generals of the South: The Foundation and Early History of The Three Kingdom's State of Wu, Faculty of Asian Studies Monographs, New series #16.* Canberra: Australian National University, 1990.

Dien, A. E., Jeffery K. Riegel, and Nancy T. Price (eds). *Chinese Archaeological Abstracts, 2, Prehistoric to Western Zhou, Monumenta Archaeologica*, Vol. 9, Institute of Archaeology. Los Angeles: University of California, 1985, (1-618).

– – *Chinese Archaeological Abstracts, 4, Post Han, Monumenta Archaeologica*, Vol 11, Institute of Archaeology. Los Angeles: University of California, 1985 (1382-2132).

Eberhard, Wolfram. *China's Minorities, Yesterday and Today.* Berkley: Wadsworth Publishing, 1982.

Fairbank, John. K. and Edwin Reischauer. *East Asia the Great Tradition.* Boston: Houghton, Mifflin Co., 1960.

Fei Hsiao Tong. *Towards a People's Anthropology.* New York: New World Press, 1981.

Fei Hsiao Tung *Peasant Life in China.* London: Routledge & Kegan, 1962.

Franklin, Ursula M. "On Bronze and Other Metals in Early China," In: *The Origins of Chinese Civilization*, ed. David N. Keightley, Berkeley: University of California Press, 1983.

Griffith, Samual B. (trans). *Sun Tzu: The Art of War*. London: Oxford, 1963.
Heberer, Thomas. *China and its Minorities, Autonomy or Assimilation*.
   Armonk, N.Y.: M.E. Sharpe, 1982.
Hersey, John. *A Single Pebble*. New York: Knopf, 1963.
Hinton, William. *Fanshen*. New York: Random House, 1966.
   – – *Hundred Day War: The Cultural Revolution at Tsinghua University*.
   New York: Monthly Review Press, 1972.
Historical Low Water Survey Team from the Changjiang River Basin
   Planning Office and the Chongqing Municipal Museum. "A Study of
   Historic Low Water Levels in the Yi[chang] to Chongqing Sector of the
   Upper Reaches of the Changjiang," *Wenwu*, 1974(8). Translated in
   Abstracts Vol 9, (86-98).
Hucker, Charles O. *China's Imperial Past*. Stanford, CA: Stanford University
   Press, 1975.
Knapp, Ronald. *Chinese Landscapes: The Village as Space*. Honolulu:
   University of Hawaii Press, 1992.
   – – *China's Vernacular Architecture*. Honolulu: University of Hawaii Press,
   1989.
   – *China's Traditional Rural Architecture*. Honolulu: Univ of Hawaii Press,
   1986.
Li Wenjie. "A Discussion of the Relationship between Daxi Culture and
   Qujialing and Yangshao Cultures," *Kaogu* 1979(2) 161-164). Translated in
   Abstracts, Vol. 9 (151-155).
   – – "Study of Classification and Chronology of the Daxi Culture," *Kaogu
   Xuebao* 1986(2). In Chinese.
Lip, Evelyn. *Chinese Geomancy*. Singapore: Times Books International, 1979.
Little, Archibald. *Through the Yangtze Gorges*. London: Sampson Low,
   Marston & Co., 1898.
Li Zehou. "Evolution of the Buddhist Divine Images as a Reflection of Chinese
   History," *Wenwu* 1978(12) (34-40). Translated in *Abstracts* Vol. 11
   (1629-1633).
Liu, Laurence. *Chinese Architecture*. London: Academy Editions, 1989.
Lo Kuanchung. *Romance of the Three Kingdoms*. New York: Pantheon Books,
   1976.
Luo Erhu. "Preliminary Study of the Cliff Tombs in Sichuan," *Kaogu Xuebao*
   1988(2) (133-167). In Chinese.
Ma Jixian. "On the Geographical Distribution of Archaeological Sites in the
   Three Gorges Region," *Jianghan Kaogu* 1988(4). In Chinese.
Ma, L. and A. Noble (eds). *The Environment: Chinese and American Views*.
   New York: Methuen and Co. Ltd., 1981.
McKenna, Richard. *The Sand Pebbles*. Greenwich, Conn: Fawcet, 1962.
Meacham, William. "On the Improbability of Austronesian Origins in South
   China," *Asian Pespectives* XXVI, No. 1, 1984 (89-106).
Meng Huaping. "A Study of the Daxi Culture," *Kaogu* 1992(4) (412ff). In
   Chinese.

Olsen, John and Sari Miller-Antonio. "The Paleolithic in Southern China," *Asian Perspectives*, Vol. 31, No. 2, 1992 (129-160).

Pannell, C. and C. Salter (eds). *China Geographer Number 12: Environment.* London: Westview Press, 1985.

- - *China Geographer Number 11: Agriculture.* London: Westview Press, 1981.

Pirazzoli-T'Serstevens, M. *Living Architecture: Chinese.* London: Macdonald, 1971.

Pocius, Gerald. *A Place to Belong.* Montreal: McGill-Queens, 1991.

Pulleyblank, E. G. "The Chinese and Their Neighbours in Prehistoric and Early Historic Times." In: *The Origins of Chinese Civilization,* ed. David N. Keightley. Berkeley: University of California Press, 1983 (411-462).

Qiao Yun and Sun Dazhang (eds). *Ancient Chinese Architecture.* Beijing: China Building Industry, 1982.

Reid, Lawrence. "Benedict's Austro-Tai Hypothesis, An Evaluation," *Asian Perspectives* XXVI(1) 1984 (19-34).

Shi Weixiang. "Two Illustrations of the Fu Tian Jing from Dunhuang's Mogao Caves," *Wenwu* 1980(9) (44-48). Translated in *Abstracts*, Vol. 11 (1632-25).

Sichuan Provincial Museum, Chongqing Municipal Museum, Fuling County Cultural Centre. "A Preliminary Report of Three Burials of the Warring States Period from Xiaotianxi, of Fuling County, Sichuan," *Wenwu* 1974(5) (61-80). Translated in *Abstracts*, Vol. 9 (876-888).

- - "Four Tombs of the Warring States Period from Xiaotianxi Site in Sichuan's Puiling County," *Kaogu* 1985(1) (14-17).

Sichuan Provincial Museum. "The Third Season of Excavation at the Daxi Site in Wushan," *Kaogu* 1981(4). In Chinese.

- - "Archaeological Survey of Wanxian County," *Sichuan Kaogu* 1990(4) (314-321). In Chinese.

Three Gorges Water Control Project Feasibility Report, Volume 8, Environment, Appendix 8G, Cultural Heritage, National Library of Canada, Ottawa, 1988.

Tian Changwu. "Investigations into Xia Culture," *Wenwu* 1981(5) (18-26). Translated in *Abstracts* No. 9 (390-396).

Waley, Arthur. *Translations from the Chinese.* New York: Knopf, 1941.

Watson, Burton. *Su Tung-P'o, Selections from a Sung Dynasty Poet.*

Werner, E. T. C. *A Dictionary of Chinese Mythology.* New York: The Julian Press, 1961.

Wong, How Man. "Hanging Coffins of the Bo People," *Archaeology,* September/October 1991 (64-67).

Wood, Bernard. "Origin and Evolution of the genus *Homo*," *Nature*, Vol. 355, 1992 (783-790).

Worcester, G. R. G. *The Junks & Sampans of the Yangtze.* Annapolis: Naval Institute, 1971.

Wu, Rukang. "Recent Advances of Chinese Paleoanthropology," *Hong Kong University Occasional Papers #2*, 1982.

Wu Rukang and John W. Olsen. *Paleoanthropology and Paleolithic Archaeology in the People's Republic of China.* Academic Press, Orlando, 1985.

Xu Zhongshu. "The Tiger-knob *Chunyu* Bell Unearthed at Xiaotianxi, Fuling, Sichuan," *Wenwu* 1974(5) (81-83). Translated in *Abstracts*, Vol. 11 (2025-2027).

Yang Qunxi. "The Cultural Sequence of Prehistoric Cultures in the Xiling Gorge Area," *Chinese Cultural Relics Newspaper*, Nov. 11, 1992. In Chinese.

Yichang Prefecture Museum and Sichuan University History Department. "The Neolithic Site at Zhongbaodao in Yichang," *Kaogu Xuebao* 1987(1) (45-97).

Yichang Museum and Archaeological Institute of Sichuan University. "Preliminary Report of Test Excavations of the Baimiao Site, Yichang County," *Kaogu* 1983(5). In Chinese.

Yichang Museum. "Summary of Archaeological Fieldwork during Preparations for the Three Gorges Dam Construction." *JHKG* 1986(1) (95ff). In Chinese.

Yin Weizhang. "An Enquiry into Erlitou Culture," *Kaogu* 1978(1) (1-4). Translated in *Abstracts*, Vol. 9 (354-357).

Zhao, Songqiao and Wei-Tang Wu. "Early Neolithic Hemudu Culture along the Hangzhou Estuary and the Origin of Domestic Paddy Rice in China," *Asian Perspectives* XXVII(I) 1986-7 (29-34).

Zhong Yuanzhao (ed). *History and Development of Ancient Chinese Architecture*. Beijing: Science Press, 1986.

INDEX

Chinese Dynasties
(pinyin and old spelling)

Xia (Hsia)	c. 21st-16th century B.C.
Shang (Shang)	c. 16th-11th century B.C.
Western Zhou (Chou)	c. 11th century-771 B.C.
Spring and Autumn Period	770-476 B.C.
Warring States Period	475-221 B.C.
Qin (Chin)	221-206 B.C.
Western Han (Han)	206 B.C.-A.D.24
Eastern Han (Han)	25-220
The Three Kingdoms	220-265
Wei (Wei)	220-265
Shu (Shu)	221-263
Wu (Wu)	222-280
Western Jin (Tsin)	265-316
Eastern Jin (Tsin)	317-420
Southern and Northern Dynasties	420-589
Song (Sung)	420-479
Qi (Chi)	479-502
Liang (Liang)	502-557
Chen (Chen)	557-589
Northern Dynasties	386-581
Northern Wei (Wei)	386-534
Eastern Wei (Wei)	534-550
Western Wei (Wei)	535-556
Northern Qi (Chi)	550-577
Northern Zhou (Chou)	557-581
Sui (Sui)	581-618
Tang (Tang)	618-907
Five Dynasties	907-960
Liao (Liao)	916-1125
Song (Sung)	960-1279
Northern Song (Sung)	960-1127
Southern Song (Sung)	1127-1279
Western Xia (Hsia)	1038-1227
Jin (Kin)	1115-1234
Yuan (Yuan)	1271-1368
Ming (Ming)	1368-1644
Qing (Ching)	1644-1911